Sara Wheeler's prize-winning books include *Terra Incognita: Travels in Antarctica* ('I cannot believe there will ever be a better book about the Antarctic' *Telegraph*), *The Magnetic North: Travels in the Arctic* ('Exceptional' *New York Times Book Review)* and *Access All Areas: Selected Writings 1990–2010.* She has also written biographies of Apsley Cherry-Garrard ('Superb' *Guardian*) and Denys Finch Hatton ('Magnificent' *Sunday Times);* and *O My America!* ('Unforgettable' *Sunday Telegraph*). The *Sunday Times* found her last book, *Mud and Stars: Travels in Russia with Pushkin and Other Geniuses of the Golden Age* 'superlative'. Sara is a fellow of the Royal Society of Literature and a contributing editor of the *Literary Review.*

A *Financial Times* and *Times Literary Supplement*
Book of the Year

'Part travelogue, part memoir, *Glowing Still* smoulders with anger about [...] the historically contested place of women travel writers ... It's also a funny and revealing account of her own journey as a writer ... colourful, deceptively capacious ... Wheeler excels at uncovering illuminating details about the people she meets' *Times Literary Supplement*

Glowing Still

A Woman's Life on the Road

Sara Wheeler

abacus
books

ABACUS

First published in Great Britain in 2023 by Abacus
This paperback edition published in Great Britain in 2024 by Abacus

1 3 5 7 9 10 8 6 4 2

Portions of the text originally appeared, in somewhat different form, in *The Economist* (on Rancho Meling in Chapter 6 and on Russian trains in Chapter 8); in the *Telegraph* (on the bayous and on Eudora Welty in Chapter 5 and on Borges and on Blancaneaux in Chapter 6); in *The Times* (on the Dalton Highway in Chapter 5); in the *Guardian* (on women's travel writing in Chapter 7); in *Vanity Fair* (on Patagonia in Chapter 6 and on China in Chapter 7); in the *Oldie* (on the death of Tolstoy in Chapter 8); in the *Literary Review* (on the singo ritual in Chapter 10); and in *Access All Areas*, 2011 (George Best in Moscow in Chapter 8). Also in broadcast form on BBC Radio 4's *From Our Own Correspondent* (on the California desert in Chapter 5, on Lucha Libre in Chapter 6, on the last mailship in Chapter 7, on Natasha's story and on Son Kul in Chapter 8) and *A Point of View* (Is that Miss or Mrs? in Chapter 2).

With the exception of those acknowledged on p 325, all photographs are the author's own

The moral right of the author has been asserted.

A CIP catalogue record for this book is available from the British Library.

Paperback ISBN 978-0-349-14510-5

Typeset in Berkeley by M Rules
Printed and bound in Great Britain by Clays Ltd, Elcograf S.p.A

Papers used by Abacus are from well-managed forests and other responsible sources.

MIX
Paper | Supporting
responsible forestry
FSC® C104740

Abacus
An imprint of
Little, Brown Book Group
Carmelite House
50 Victoria Embankment
London EC4Y 0DZ

An Hachette UK Company
www.hachette.co.uk

www.littlebrown.co.uk

For Chris Coles
Celebrating forty years of friendship

Contents

I do not want to be good ... I wish to be hell on wheels, or dead.

MARTHA GELLHORN

Introduction

I n *Cameroon with Egbert*, Dervla Murphy encounters a curious observer. 'Are you a man or a woman?' he asks. In response, Murphy lifts her jumper to show her tits. *Glowing Still: A Woman's Life on the Road* records waystations on the female travel writer's journey, launching at Nubility and voyaging, via children, to the welcoming port of Invisibility. (We will leave Immobility for the next volume.)

I came of age, legally, between LSD and the SDP. Role models were scarce in the travel-writing game. The seventies had launched a golden age, but few women set sail. On an otherwise male-only Antarctic base I stowed used tampons in the pocket of my parka, where they froze into objects resembling stock cubes. As the decades unfurled – Pole to Pole, via Poland – emotional trajectories lent meaning, if not coherence, beyond the practicalities of the story.

Urgent concerns have lately overpowered any thought of 'meaning'. When I started my so-called career it was good to travel – even morally good. But three horsewomen of the apocalypse write the contemporary agenda: Covid, voice appropriation and the vital need to prioritise the environment. It's flat-out bad to take off, almost wicked – you destroy the planet, kill old ladies and use your first-world white power to subordinate the Global South by writing stories that belong to writers of those lands.

And who is that kicking up dust in the rear? It is a fourth horse-woman, carrying the message *We've Been Everywhere*.

What are we to make of this, and where are we heading from here?

As for the few role models which did exist: I carried them in my head like the toothbrush in the sponge bag. Reading and travelling are parallel journeys. I am in the debt of the other women who not only set sail but also unsparingly observed the world that spins within each self. They set me off, and I turned to them again when I decided to look back at my travelling life for this book. Then when I took battered notebooks from their shelves to remind myself of what I had done, I discovered that I had often left out the best bits. Being sixty brings its own freedoms, one among them a disregard for what anyone thinks, so in these pages I have set down what I consider important, or revealing, or funny. Sometimes I added what neither my books nor the notebooks say – great things might have occurred at the Pyramids one afternoon, but all I remember is crying in a single bed because a sweetheart cut me in a letter. It's the emotional residue that sticks. There were *amours* on the road, and I mostly chose not to include them in the books. Anyway once they had expired like a passport, I learned of a more serene peace: that of the double bed with nobody else in it. I hope this book conveys something of that trajectory from the chaise longue. Beryl Bainbridge once told me in the back of a taxi she was glad 'that part of my life' was over. I was twenty-nine years her junior and thought, I can't imagine that happening, and when it does, it won't be living, it will be something else. But it has happened, and it's fabulous. Beryl and I spent a lot of time in the back of taxis. Once we had a father-and-son crush going on. She was kind to me when I was starting out, putting me up as a Fellow of the Royal Society of Literature, giving me a generous quote for one of my early books and picking another as her book of the year. I miss her. (Both the father and the son crush were unrequited, unreciprocated, pathetic in a good way.)

Despite the hot breath of the horsewomen's steeds, writers do still get out there. One wonders though how the professional landscape has changed since I started out all those decades ago. Testosterone has always pervaded the genre, but over the course of my career the 'I've-Got-A-Big-One' school has ceded ground. Early on I remember getting lost in the bowels of Broadcasting House with a leading male explorer after a late-night Radio 3 programme on bold travels. After we had to ring reception for assistance to find our way out, he said, 'Don't tell anyone.'

Women, less interested in 'conquering', have pioneered a kind of creative non-fiction that suits the travel genre. I prefer it to the blokeish business of seeing how dead you can get. It notices more. Martha Gellhorn's blend of reportage and imagination ensnared me when I was barely out of my teens, and her preferred form has come of age in my working life. Not only do I think Gellhorn is a marvellous writer – at her best, one of the best – but I also identify with Gellhorn the woman. 'The open road,' she wrote, was 'my first, oldest and strongest love.' The question I am most often asked is, 'Weren't you afraid?' Of predators, of the hidden crevasse, of the bitter loneliness? Perhaps, but if I were I can't remember it. The John Lewis curtain department terrifies me most, and when I am there I think of Gellhorn. She lived from 1908 to 1998 and was writing the fighting for six decades, and although each conflict was different, her message remained the same: 'There is neither victory nor defeat; there is only catastrophe.'

Martin Amis said when a writer reaches sixty he's got this whole new empire that didn't exist before – the past. Another country indeed. A trip appears differently in the light of reflection and experience. This book, at least to a certain extent, looks inwards where the others looked out. I see now what I was looking for on a particular journey, whether I found it and if it had value. And anyway I am writing this introduction at journey's end, looking at dhows sailing in the monsoon breeze off the coast of Zanzibar. Everyone calls me mama here, as white hair is like semaphore. That's what it's come to, and I'm glad.

A thread of memoir runs through the pages of this my tenth book. What made this travel writer? My ancestral roots descend deep on both sides into the St George district of east Bristol. But after my twenty-one-year-old mother produced me at Bristol Maternity Hospital on the first day of spring in 1961, she brought me home to a flat above my paternal grandfather's small building yard in Westbury Park, a neighbourhood of mainly unbombed Victorian housing stock in the northwest of the city, between Henleaze, Redland and the Downs. The war had ended only sixteen years previously. An anti-aircraft shell had fallen onto Auntie Jean's bed in Cowper Street (she was my father's sister). On another night, 24 November 1940, my grandfather, air-raid warden Wilf Wheeler, came to fetch my eight-year-old dad from the Morrison shelter in the garden (the Council had distributed the shelters free to households with an income below £350 a year). 'Look at that,' said Grandpa, leading his boy into the street. The city centre was on fire, and a corona over Castle Street cast a weak orange glow into starless heights. The following year, on Good Friday, my great-grandfather, a policeman, helped push the last tram ever to run in Bristol from Old Market to the Kingswood depot. A twenty-five-kilo HE bomb had fallen on St Philip's Bridge next to the Tramways generating centre, severing the power supply. This was not yet history when I was born. It was recent. Everything I saw as an adult, I tried to see in context.

I know now that events twenty years ago seem to have occurred last weekend. The war was still close in 1961 as my parents and aunts and uncles watched high-rises rear over their bombed-out city from the top decks of the buses which had replaced the trams. The Edwardian Harold Macmillan was prime minister. That year, he told the *Daily Mail*, 'We've got it good.' (Four years earlier he had delivered the 'never-had-it-so-good' speech in Bedford, so the phrase must have lingered in his mind.

Either that or his speechwriters had run out of steam.) People were eating a fifth more than they had in 1950. 'Let's keep it good,' Macmillan concluded. The day my mother told her parents that she was expecting me – their first grandchild – the judge in the *Lady Chatterley* case handed down his verdict. A new world beckoned, though I am not convinced any Wheelers understood the gesture inviting them into it.

My maternal grandparents, with whom I spent a great deal of time in my early years, had a working outdoor lav at the end of the garden and on it Nana enjoyed long conversations with Mrs Somerville on one side and Mrs Ward on the other (they stuck to surnames till the end). The first floor of the two-up two-down terrace already enjoyed an indoor bathroom in 1961, but the fresh-air facility remained in service for some years – I remember torn squares of newspaper speared on a nail. I suspect residents were both anxious not to wear out the gleaming porcelain and suspicious of the reliability of indoor plumbing. At any rate we were fortunate: in Manchester in 1961, nearly a fifth of households did not have exclusive use of a hot-water tap and in Birmingham 15 per cent did not have their own WC. Not everyone 'had it good'. The Standardised Mortality Ratio (SMR) decreased between 1931 and 1961, meaning Britons were living longer, but the mortality rate differential between social classes (ranked from I to V) *widened* during that period. *Some* had it good.

Nan and Pop had bought the house in St George's Clovelly Road in 1932 when they married. I can see Pop in the back room, holding pikelets to the fire on the toasting fork now hanging on my office wall. Sometimes as a very small child I tottered down the stairs with him first thing to light the grate. Once the flames took, he set to bandaging both legs from the knee down, the only salve for his varicose veins. Pop was a paint salesman, but most of my family worked at the Wills tobacco factories in Bedminster and Knowle. Auntie Ruby was a tobacco-leaf stripper.

Uncle George ended up, after two years being tortured as a POW in the Far East, on the production line at Fry's Somerdale plant in Keynsham, a factory which had its own railway branch line. He brought home imperfect bars of Five Boys, the first chocolate I ever tasted. Both Wills and Fry's, everyone agreed, were good employers, and the industrial relations paralysing the rest of the country impinged little on our corner of the southwest. But there was anxiety. In the spring of 1961, 11,500 Bristolians signed a petition backed by the *Evening Post* against the piecemeal development of the heavily bombed Wine Street area where the Broadmead shopping centre had already partly risen. Auntie Ruby, the tobacco stripper, told me before she died three years ago aged ninety-eight that planners did as much damage as the Luftwaffe. It was, up and down the land, the era of slum clearances and vertical development. People feared being uprooted and transplanted to housing estates. The year my parents married, the MP for Bristol South had made a speech in the House reporting the conflicts and anxieties around slum clearances in his constituency; the question he heard most frequently on the doorstep was, 'Can you tell me when this street is coming down?' Bristol had a poor post-war planning record. The year I was born, a wrecking ball destroyed the arcade of a nine-hundred-year-old Norman house in Small Street in order to widen a courtroom. The following year, planners authorised the destruction of the fifteenth-century St Augustine the Less to make way for a hotel extension. Three fourteen-storey blocks replaced the Georgian villas of Kingsdown. In August 1961, John Betjeman told the *Daily Mail* that all developers should be put in prison 'for the crime of murdering our souls'.

My maternal tribe were Methodists. On Good Friday the men carried a wooden cross over Troopers Hill, where John Wesley had preached in 1738, and sang his brother Charles Wesley's 'O Love Divine, What Hast Thou Done!' in the open air. Ironically, given my later history, I was in the retinue of the Temperance

Queen at Bethel, the Methodist church on Clouds Hill Road (now the Lifestyle Fitness Centre). The queen, the impossibly glamorous Lorraine Marsh, processed with her retinue along Hudds Vale Road and I carried a red corduroy muffler of astonishing sophistication (Nana had made it). Bethel was a place where women wore hats and sang in queer RP diction. The walls were bare, Methodist-style. I could no more have dreamed of the ornate Catholic aesthetic I came to love as an adult than I could have imagined Lorraine shagging the vicar's son, which, it turned out, she was.

Nana, born in 1912, was the policeman's daughter. She harboured a lifelong fear of pubs, remembering Saturday-night knocks on the door and a child's voice asking her father, 'Can you come, our dad's hitting our mum?' Nana was a sparrow of a woman and after Pop died she kept her father's truncheon by the door to beat off intruders. I am looking at it now, nestling up to the innocent toasting fork, scratched and worn out, as she was in the end, and, as she had Dupuytren's contracture (as do I), with her little and ring fingers bent into a claw.

I began at my state primary school in 1966. Many teachers remembered prize-giving on 19 July 1940, when an air-raid warning interrupted proceedings for an hour and three-quarters and the Bishop couldn't deliver his address, I imagine to widespread delight. The Luftflotte bombers were heading for the aircraft works at Filton. Nobody's father was unemployed in my class and nobody was foreign. When my classmate Sally Thomas's father, an electrician, died by electrocution, the event lodged powerfully in our imaginations as we were boomers who had not seen bombs fall. Martin Burt though lived in one of the many prefabs erected as temporary accommodation when German munitions had destroyed Bristol homes. When I went to play in the Burt prefab in Henleaze, Martin's mother gave me a Fox's Glacier Mint. The prefabs turned out to be indestructible, outliving all their residents.

The television and the motor car rose to prominence in ordinary people's lives in my early years. Five and a half million cars patrolled the streets of Britain in 1960, compared with 2.3 million in 1950. We did not have a car, but we did have a black-and-white TV, and my earliest memory is standing in front of the screen on 30 July 1966 as Gottfried Dienst blew the full-time whistle at Wembley. As my mother was working, my father, a fanatical sports fan, was looking after my brother and me. Dad had promised to take us to the Downs at the end of the match. I had just learned to tell the time. At the ninety-minute break, I stood with my back to the set. Might I have said, *I think it's all over, Dad*? As both teams prepared for extra time, he offered bribes which we have regrettably forgotten, and the three of us sat it out till it really was all over, and then went to the Downs. We were a football family, my mother's family the Gas (the Rovers, whose stadium abutted the gasworks) and dad's the Robins (the City, who wore red). Dad had a season ticket to Ashton Gate and took me with him, pushing me under the turnstile. On alternate Saturdays we went to see City Reserves, as going to an away game would have been like travelling to Mars.

Television battered the pub trade: almost one pub a day closed in Britain in the year of my birth, and about the same number of cinemas drew their velvet drapes for the same reason. The twenty-four-year-old Tom Stoppard, a journalist in Bristol in 1961, recorded a visit to the Kingswood Odeon, a much-loved haunt of my mother's which had just transmogrified into a ten-pin bowling alley, according to Stoppard 'Thumping with loud, bright, interior-decorated, Tannoy-guided, muzak-placated, electronically operated life'. I realise now that my parents could have been throwing balls at the same pins in the same room as the author of *Rosencrantz and Guildenstern Are Dead*.

Yuri Gagarin and Valentina Tereshkova, the Cuban Missile Crisis, the grassy knoll, Mỹ Lai, Biafra, *les évènements*, the moon landing (*a phallic triumph, an adventure/it would not have occurred to women/to think worth while*),[1] striped toothpaste. My parents did not vote for Harold Wilson in 1964, but the majority of the electorate did, with the result that he became the first Labour prime minister since Attlee. Aberfan and the Torrey Canyon percolated into domestic discourse as both disasters happened in places we knew.

By the time I was ten I had absorbed the fact that Britain considered itself a top-table power. I did not know the motor industry was already losing ground to West German competition, as were textile and machine-tool manufacturing. Shipbuilding too was collapsing. The Suez Crisis rang the alarm for fuel supplies. There was a sense that we were falling behind, but I do not think it filtered through to the working classes of St George. Further afield, the empire was crumbling. Supermac had delivered his Winds of Change speech in February 1960 ('Whether we like it or not, this growth of national consciousness is a political fact'). Many of the countries I was to visit as an adult were casting off colonial shackles. Bristol absorbed immigrants, but did not welcome them. 'A Black friend of mine,' Stoppard reported in the *Western Daily Press* three weeks after I was born, 'applied for a job as a conductor at the Bristol Omnibus Company.' No vacancies. 'Five minutes later I applied for the same job. I was told I could come back after the holidays when there would be vacancies.' Two years later the same nationalised company put racial discrimination, and Bristol, on the national news when it publicly endorsed its refusal to hire drivers and conductors of colour.[2] The day after his bus report, Stoppard revealed that an accommodation agency had told him that just 1 per cent of Bristol's landlords would accept a non-white tenant, and another agency said that '25 per cent of landlords stipulated no coloureds without being asked'. Half

1. W.H. Auden, 'Moon Landing'.
2. The behaviour of the transport unions degrades the already degrading boycott story – a saga my parents remember. The TGWU Bristol chapter said that 'if one Black man steps on the platform as a conductor, every wheel will stop'.

the other property owners 'would not accept a coloured tenant even if one came along'. Stoppard concluded his investigation, 'It is never the landlord who doesn't like coloured people. It's the other people in the house, or the other people in the street. No one you meet is guilty.' At the Tory Party conference in 1962, a swivel-eyed backbencher, ancestor of today's troop, distributed copies of a letter he had had published in the *Telegraph* titled IMMIGRATION LUNACY. Looking back, I note that my life has played out not in an arc but following a straight line that started low and descended. Britain's relationship with Europe *was* an arc in my lifetime: one that finished on the same axis it had begun. On 31 July 1962 Macmillan announced we were applying to join the EEC, known as the Common Market.

The most handsome row of houses in Westbury Park stretched opposite the Downs on the site of a medieval monastery. This was St Christopher's School, founded in the rubble of war for the eleven-plus education of what were then called mentally handicapped children. Mathew, my only sibling, attended the school for many years and I got to know it about as well as one knows any place as a child. Matt is profoundly intellectually disabled, with severe behavioural problems. He has never had a diagnosis. They don't know what made him atypical. He has no physical problems. Perhaps the fact that he cannot read his own name made me a reader later. We looked alike. Our mother did not want Matt to board at St Christopher's but she could not cope, so he did. I am the sister of an only child. Many of Matt's peers were in the care system. They had no homes. We picked Matt up from his boarding house most Fridays and at the end of term. Every December, pale faces looked at us, colourless (at best) weeks at school lying ahead while we opened presents. These experiences lay down a sediment in you.

As the late sixties became the early seventies, memories swim into focus, though I cannot recall any political issue ever being discussed at the tea table, tea being the evening meal, consumed

at 5.30. Except, that is, capital punishment, which everyone wanted back after parliament abolished it (with caveats) in 1965. Every member of the family was a blue-collar Tory. They all read the *Daily Express*. There was no unconscious bias. It was all conscious. We didn't like anyone who wasn't like us.[1]

Decimal currency arrived in 1971, just before I won the school prize for the best logbook following an epically thrilling school excursion to Ostend on which family-sized glass bottles of Coca-Cola stood on the hotel tables and we could help ourselves to as much as we wanted. The prize was a postal order. But what on earth was *25 pence*, and what could it buy?

Simone de Beauvoir's *The Second Sex* had appeared twelve years before I was born and sold twenty-two thousand copies in a week. Feminism had not however made its mark in Westbury

---

1. The reader might be surprised to learn, in the light of this parochial childhood, that in 1971 I met George Best in Moscow. This was an anomalous event and one which does not fit into this brief introduction, which situates the small, homogenous world whence I sprang. But you will read about it in due course, later in the book.

Park. All the grown women I knew were wives and wives alone, bar a few spinsters like Auntie Ruby, the stripper. 'The domestic labours that fell to her lot,' de Beauvoir wrote of twentieth-century woman, 'because they were reconcilable with the cares of maternity imprisoned her in repetition and immanence ...' A 2012 academic publication estimated that in 1961 the working-class woman in Britain spent seven and a half hours a day engaged in housework. Betty Friedan cited de Beauvoir as an influence on *The Feminine Mystique*, which appeared in 1963 and which in turn heavily influenced (at the very least) the Second Wave, of which I caught the break. I was the first of my family to continue education beyond the age of sixteen (only my mother reached even that milestone). Having benefited from the best state-financed education any girl ever had, and on account of it, I progressed to the realisation in middle age that, deep down, I'm shallow.

Shallow or not, seven and a half hours is a lot.

Working-class life was on the move in some quarters, edging towards middle-class values. This manifested itself in wine-making kits purchased in Boots on Gloucester Road. In 1971 my parents even joined the ranks of home ownership. Our Bishopston bungalow was like Xanadu. It had a garden with an apple tree, a hall with a tiled floor and for the first time I had my own bedroom. It had a Ladderax shelving system which the previous residents had tried but failed to wrench off the wall, and a window onto a side path which was to function as an escape route in my early teenage years. On our first week in the bungalow Matt pushed me into the new bath from a stool on which I was standing. My head cracked open, and I had stitches without anaesthetic at Cossham Hospital.

The environmental movement had not begun to move, at least in Westbury Park. Rachel Carson, fugleman for the apocalypse to come, published *Silent Spring* in 1962. The hazards of industrialised farming were in the public domain but the public were not aware of them. The insecticide dieldrin, a bioaccumulative endocrine disrupter, had been withdrawn from sale in the UK but not banned. Carson, whose book was a bestseller (it was the birds which were silent), revealed that pesticides were killing the planet. In the immortal *The Peregrine*, also published in the sixties, J. A. Baker wrote that toxic chemicals had produced, 'a dying world, like Mars, but glowing still'. These two movements, feminism and environmentalism, were the lenses through which I saw the world when I started out on my actual journeys.

Of the literary world we also knew nothing. I am not sure anyone sensed Godot might be imminent. But as Estragon says, 'There is no lack of void.' Today I read in cultural histories that *The New Poetry*, which Penguin brought out in April 1962, was 'influential', but nobody in Westbury Park, to my knowledge, raised a concern that the anthology *contained no women poets*. There were no books by women in the flat above the builder but none by men either, except a yellow hardback novel without a dust jacket by Howard Spring. I can't remember any children's books; I don't remember any at my primary school either, but there must have been some. My father used to sing Matt and me to sleep with 'Daisy, Daisy'. When he got to the end, I would pipe up, 'Again!' I've just telephoned him to check my memory is sound on this point.

This, then, was my *Kindschaft*, the fused experience of childhood and landscape. Travel was not on anyone's minds where I came from. Britons took 4 million holidays abroad in 1961. None of them lived in Westbury Park. Over my first decade, the figure rose to 7 million. The adult voices I heard in my early years expressed suspicion of Abroad. It was obvious to me that everyone considered British people superior to any other kind. My parents' friends used to say, 'over there' to mean anywhere that wasn't Britain. Was that what made me a traveller? I don't think so. At least, I didn't seek in my writing to show that foreigners weren't inferior and therefore prove my kinsmen wrong. But perhaps it did make me go, for another reason. Because I wanted to dilute the people among whom I had grown up. At any rate, responding to an otherwise undetected frequency, I left.

Looking back, I see there was something cuspy about my juvenile world. As I approached the respectable age of ten I absorbed some notion of modernity even as Pop pulled a chair across the back room in Clovelly Road to retrieve cash from the pelmet, a safer place to store it than the bank. (They did though trust the Wiltshire Friendly Society, from which I received each year a Christmas gift of a half crown in a small brown envelope.) We were shifting towards a cultural centre, and started going to the Bristol Old Vic when my mother got a job selling ice creams and we received free tickets. The authenticity of a working-class identity was slipping away. Did rising affluence usher in revised values? Alan Bennett said, 'I wish the sixties hadn't happened, because that was when avarice and stupidity got to the wheel of the bulldozer.'

When I decided to bring the strands of my travelling life together in this book I moved the kitchen steps to my office in order to pull down my notebooks; they had the names of places written on the spine, sometimes in Tippex. Some destinations had notebooks to themselves (*St Helena*) and some made odd bedfellows (*Fiji, Malawi*). I was amazed at the gathered detail, and had forgotten much, though the friable pages brought it back – I saw again the colour of the edging on the nurse-nuns' saris at Mother Teresa's mission in Calcutta; felt the pre-dawn cold of the Gare du Nord and the clubbing heat of an outback bar north of the Ningaloo Reef in Western Australia with a gutter at the foot so men could piss without moving. I had a USGS map in my pack once in the far south and when I spread it out on the ice I saw I had reached a line drawn with a ruler marked *limit of compilation*. I had reached the end of the map. Just before the line, geochemists working in the region had christened an espe-cially pointy mountain the Doesn't Matterhorn. You could name

anything there then: my only experience of eighteenth-century life on the road, excepting scabies and malaria. Sensations rose like milk to the boil when I re-read the tattered notebooks, and I found it tender to be in the company of my younger self. I felt I saw her clearly, though I know enough about memory to be aware of the fallacy of that statement. In some ways, as I started to write, I met myself coming back. At any rate, we will start, in Chapter 1, far from Westbury Park. But I remember, and so will you, that it was always there.

1

At Least 300

Antarctica

Because she is woman, the girl knows that the sea
and the Poles, a thousand adventures, a thousand
joys are forbidden to her: she is born on the
wrong side.

SIMONE DE BEAUVOIR, The Second Sex

They had set up an ice hole for the bathroom, placing a wooden box with a polystyrene seat over it and fencing it round with a windbreaker. It was the best-appointed lavatorial facility the world had ever seen, facing the corrugated cliffs of the Mackay Glacier and, to one side, thousands of kilometres of ice. But one day a geologist slithered across camp with his windpants round his knees. 'Crikey, Mike!' I yelled. 'What's wrong?' He had been enjoying the view from the khazi when a seal came up the ice hole. Hot fishy breath, apparently.

The five-strong benthic geology team working on the frozen Ross Sea were sending an ROV (remotely operated vehicle) to the bottom to capture images: debris dispersal yields clues to the speed of glacier erosion, a factor in climate variability. I

looked out from my tent – I left the flap open during the bright
nights – at creamy ice walls reflecting saffron sun, or an arc
of lenticular clouds, or the plumes of Mount Erebus marbling
the sky. Like Christopher Isherwood arriving in Venice, I was
astonished to think that the Antarctic had been there every day
of my life. The group's field assistant, a New Zealander impres-
sively bearded even in a barbate land, had found among camp
paraphernalia a board to place under my sleeping mat to prevent
Eurydicean descent into the ice. It was a happy camp. The three

junior geologists, all with sunburned faces and white eyelids, were intrinsically decent people, but inevitably took their cue from the project leader – their boss, effectively. This top-down cultural influence poisoned my residence in the Antarctic six months later far from Mackay. I want to show here at the outset that the sailing has not been plain. De Beauvoir was right, in the part about the sea and the Poles and the thousand joys – you can still feel 'born on the wrong side'. In the course of the painful experience I will describe later in the chapter, I began to think about the actual position and the perceived role of women on the road. *Terra Incognita* was my third book, and I was thirty-four. But, somehow, until then I had just got on with it.

I spent seven months in the Antarctic. Records do not reveal a particularly cold year, but I once experienced minus 115 with wind chill. The sun often shone. At the Pole itself, some days I strolled around without a jacket. Breath came short, as the Earth's atmosphere is at its most shallow at the two axes of a rotating planet. I was walking on a layer of ice a third as high as Everest (the altitude is 2,850 metres) and the combination of elevation and shallow atmosphere means the body receives half its normal oxygen supply. At another camp, when it was very cold but windless, and after a pair of emperor penguins had visited our row of pup tents, a colleague threw boiling tea in the air and it froze before it hit the ice, tinkling modestly as it shattered. A field camp always involves donkey-work and I took on the worst jobs as a means of ingratiating myself. For long hours I sat on a box on a frozen lake spooling hundreds of metres of tubing into the depths, or glued my eyes to a gauge on a Twin Otter dashboard to ensure the pilot maintained a fixed altitude, as a radar was sucking data from the ice sheet below as we flew. (This was indeed a terrible job.) Before take-off the skipper had warmed the engine with a hairdryer powered by a small generator. We were close to the Sky-Hi Nunataks in Palmer Land at the base of the Antarctic Peninsula, the long continental finger tapering towards South America. It was the end of a poor season and the scientists had gone home, leaving eight support staff. The cook tent was permanently flooded with brown ice-water and when I arrived the men were passing round a catering tin of peaches with an upright spoon in it, penguin feathers stuck to the label. Field-camp tucker had often been frozen for a long period, not in a freezer but in an Antarctic storage hut. On a tub of Parmesan printed with the words, 'matured three months', a wag had written with a Sharpie, FROZEN TEN YEARS. I had though actually seen a fridge in an Antarctic camp. Powered by a gas canister, it stored samples from a lake in the Dry Valleys to ensure they didn't freeze. It was a fridge to stop things getting cold, a fantastical inversion of logic which was not the only Wonderlandish feature of the seventh continent. 'How long is for ever?' I once overheard an astrophysics grad student ask another as they washed up (the scientists returned regularly to the debate over whether scrubbing dishes in water melted with fuel was an ecological improvement over burning paper plates). 'Sometimes just one second,' the other man replied, and I was never sure if he was quoting *Alice* or replying seriously, as something happened at that moment to divert my attention; I expect it was another penguin. The birds have no predators out of the water, and they never tired of their research into bipedal behaviour.

I enjoyed the coastal camps, learning about krill and creatures that dwell deep in the dark, cold water (the team studying the latter called themselves the Bottom Pickers). I consorted amicably with cryptoendolithic microbial communities in the Convoy Range on the continental edge where baroque landscapes of rich red and gold glitter over the 2 per cent of the continent not permanently shrouded in ice. For a period I was the American government's Writer-in-Residence, and during that time I camped close to the Erebus Glacier Tongue for six weeks with the Artist-in-Residence. Her job was harder, as her tubes of paint froze. These were the August days when the sun rose, and we two were the only human beings camping out on land one and a half times the size of the continental US. In September engorged female Weddell seals flopped out of ice holes on the frozen Southern Ocean and pupped ink-dark alongside us. Weddell milk has the highest fat content of any mammals' and the babies gained two kilograms a day before our eyes. It was like watching dough rise. Sometimes we saw the Southern Lights that month. A rippling fluorescent curtain swished from east to west in radiant lilacs, greens and pinks, and from it emerged an evanescence of strands which stretched into ribbon streamers and danced. The ends of the ribbons frayed when they curved, forming a prismatic fringe of tender cobalt. At other times bars of light marched in close-packed succession, transforming themselves at a set point into solid white lances or gleaming rods of glass. Colour saturates my memories of what is supposed to be a colourless land. The presence of the painter and her tubes led to a habit, one that persists to this day, of using the names of paint colours to describe things. Naples yellow? Plenty of that in my pages. Manganese blue hue – yay!

I slept on Captain Scott's bunk in the Cape Evans hut, the one to which he never returned after setting out for the Pole in 1911. (When he saw the Norwegian flag flapping at ninety south, Scott wrote in his journal, 'The worst has happened.') I say 'slept': it was more a case of maintaining a heartbeat in the unforgiving cold. Katabatic winds cast stones at the window. I looked up at the quilts Scott's men had stitched with pockets of seaweed to insulate the walls. When the others in the hut were waiting for the polar party to return in the long months of the austral winter, they said that when they heard knocks on the panes, they thought it was them, back at last. But it never was. There is much that is heroic about Scott. 'I shall never fit in my round hole,' he wrote home to his wife. I like a doubter.

In a signing queue years later for a book set far from the Antarctic, a member of the Scott clan waited patiently in order to say, 'We disapprove of you sleeping on the bunk. What if everyone did it?' As I signed her copy of an old hardback she had brought with her, I thought but didn't say that nobody else was there – not for several thousand miles.

At Cape Royds I found in Shackleton's hut unrusted tins of sardines, Huntley & Palmers biscuit crates and stacks of Belmont stearine candles. For the Boss, Antarctica was always a metaphor as well as an explorer's dream: he once wrote that 'We all have our own White South'. A colleague noted in his diary that, for Sir Ernest, 'Antarctica did not exist. It was the inner, not the outer world that engrossed him.' Was that me? During that visit Antarctica took up position in the landscape of my imagination as the place where everything is possible and nothing spoils. In the biological haiku of the interior, thousands of kilometres of ice bat solar energy back up into the sky – a perfect metaphor for a spiritual powerhouse. I miss the days of mapless land and nameless places.

As a guest of the US National Science Foundation I camped mainly with Americans, but other nationals were frequently on the team, contributing expertise and sometimes, through their home institutions, sharing the burden of research finance, fearsome in the polar regions. Thirty Italians and their three Kiwi pilots were working at Terra Nova Bay near the Drygalski Ice Tongue in Victoria Land. The red Squirrels were as small as a helicopter can be, and had quarterlight windows like a VW Beetle which one pilot opened as we cruised above the ice in order to tap ash from his Marlboro. On a prior visit to King George Island with the Chilean Antarctic programme I had climbed nunataks and consumed reconstituted, and thankfully unidentifiable, rations with polar South Americans, with Russians at their Bellingshausen base, and with Chinese scientists at Great Wall, which a two-metre fibreglass panda sentinel guarded a considerable distance beyond the range of its habitat. Happy memories, and when I climbed down the metal steps of a Dash 7 outside

Rothera station on the Antarctic Peninsula I was among my own kind in the Big White for the first time. I had approached the British Antarctic Survey asking if they might take a writer south. They wrote back to say yes. Of course they didn't, but that is how people imagine it happened. 'You're so lucky,' they say. In reality the Cambridge-based Survey and I had spent two years engaged in talks and negotiation while I was involved in simultaneous discussions with its US counterpart, the Virginia-based National Science Foundation, and with publishers on both sides of the Atlantic. In the end all four said, 'We will if they do.' I had to earn a living like everyone else during those tricky twenty-four months, with no seals in view from the bathroom. But now I had flown 1,860 kilometres to Rothera from the Falklands, having reached Port Stanley the day before on an RAF VC10 from Brize Norton, via Ascension Island. The De Havilland turboprop Dash I boarded at Port Stanley after a fortifying night at the Upland Goose was a resupply aircraft carrying pallets of food and science equipment as well as two other passengers, one the manager of a building project under way behind the main base station. For five hours we floated over the Southern Ocean higher than the alba- trosses diving on the wind above Melville's 'effulgent Antarctic Sea', past the tip of South America and on south, skimming peri- Antarctic islands and the west coast of the Peninsula. Then we reached the highly glaciated Adelaide Island and the research station crouching close to sea frozen to a depth of two metres.

The Dash door opened and the now familiar gust of freez- ing air freighted with diamond dust powered into the plane; I descended the fold-down ladder; and stood there while the two pilots began to unload and the other passengers greeted hooded figures who had walked over to meet them. Minutes passed. I could see the main base building on the side of the airstrip, a low, long structure with a green roof and a short flight of front steps leading to a door. To my right, a tabular iceberg rippled in a blue haze.

I couldn't stand there until something happened, in case nothing did. I walked over to the building, went up the front steps, pushed open the door, which was like that of a walk-in freezer – again, this was familiar – and entered a corridor through which hairy men bustled. Nobody spoke to me. Nobody told me where I might find my pitroom, or indeed if I had been allocated one. Was I to sleep outside? I was alone among my countrymen at the end of the world. Where was Melville now?

It turned out – and it took me a long time to understand this – that certain personnel within the institution resented the presence of an outsider. Especially a woman. And a non-scientist. 'I thought how unpleasant it is to be locked out,' Virginia Woolf wrote in *A Room of One's Own*. My mind's eye looked down the wrong end of a telescope at all the women through history who had been shut out from places they wanted to go. Twenty-five men were at Rothera when I arrived and a misogynistic culture had set as solid as the bergs in the bay. When I entered the galley, people cracked hate jokes. Here's one. 'Why do women have periods?' 'Because they deserve them.' Fucking hilarious! A field assistant – let's call him Derek, because that was his name – made remarks about my weight. During the weekly Film Night, when I again entered the multi-purpose galley, someone switched the film to the *Viz* comic's *Fat Slags*, fast-forwarding to the scene in which a character noisily scratches her fanny. I tried to look out of the picture windows and focus on the growlers at the edge of the water, or the airstrip gravel. Sometimes the light above Reptile Ridge was buttery. I reached for redemption in the landscape, writing in my diary, *There can be nothing else on earth like an Antarctic summer's day, and it lasts all night.* Perhaps that is what Shackleton did when the inner world grew cold. But hourly humiliations came my way. Oh, it was horrific. When I came to write *Terra Incognita* I left out the abuse out of loyalty to the BAS director in Cambridge who had supported my project. But I'm glad to get it out of the way now. Why the brittle mask? was all I asked in that book. At least it became a bestseller, which annoyed them.

I learned on the Peninsula that misogyny stalks the female traveller's world, and I have learned it many times over in the subsequent decades. Are British men worst? Yes, I think they probably are, though white South Africans mount robust competition. Why? Is it an empire-remnant in which men think they did all the exploring? Or is it something more profound? Taslima Nasrin, the Bangladeshi-Swedish writer and human-rights activist who is about my age, writes this: 'Women are oppressed in the east, in the west, in the south, in the north,' and she continues, 'Women are oppressed inside, outside the home.' In March 2007 the All India Muslim Personal Law Board (Jadeed) offered 500,000 rupees for Nasrin's beheading. So it's not just British men; of course it isn't. I think, though, besides the issue of leadership influence, the Antarctic functions as a Petri dish. The isolation cuts people off from the corrective forces of civilisation. The British are the worst as their programme has deeper roots, nourished in the era of John Rymill, whom you can see on flickering footage with the team he led on the British Graham Land Expedition in the thirties, skiing down glaciers like Greek gods. Rymill and the others had themselves been raised on the figure of Captain Scott, by then Galahad of the snows. Other nations had no Antarctic mythology to foster chauvinistic ideals. On the other hand (I am going round in circles now), according to Stephen Martin's *A History of Antarctica*, male staff on a non-British station 'left pornographic material at the dinner place of a female staff member on seventeen consecutive nights'.

An early lesson in anxiety management when far from emotional base camp, in this case on an actual base camp. How should one learn to cultivate the positive outlook over the negative, on the road and off it? This was a life's work. I had written out a line from Maud Parrish at the front of my first Antarctic notebook. She was a teenage runaway whose travel book *Nine*

Pounds of Luggage appeared in the US in 1939. In the depths of despair and Dawson City, Parrish wrote, 'Some see life in black-and-white; others – they're the lucky ones – in old-gold hues.' (Or was it burnt umber?)

Science fostered prejudice of its own. Men thought they owned the ice (but they didn't), and some men felt that scientists thought *they* owned it. I was an Ed Hillary fan, my admiration based not just on his achievements on Everest and in the Antarctic – though they would have been enough – but also on his modest demeanour. This was not a mere slayer of mammoths. He was, as a Kiwi, outside the fearful class system exemplified then and now by the Royal Geographical Society, the institution behind so many expeditions. I had tea with Hillary in Auckland once. They say you should not meet your heroes, but Sir Ed did not disappoint. He rose in my estimation. He was quiet and funny. Later, when I was trying to make sense of my polar months and commit them to paper, I wrote to ask about the ownership business. He sent a handwritten reply stating that in his experience scientists felt they owned the Antarctic and that mere mortals such as explorers had no right to be there. They really thought it was their own white south.

I've started with this issue in order to situate *Glowing Still* in the long story of women on the road, itself part of the longer story of women parked in the sidings, which is the one of all women, in all times, everywhere. With or without a pram or an ice axe in the hall. Why does any woman ever set out for the thousand adventures?

The first female travel writer to have left a surviving record voyaged to deepen her faith, or that's what she said she did. In the fourth century CE a woman we know as Egeria set off for the holy lands of her Christian religion, stayed away for three years and composed a long letter to girlfriends describing her trip. If travel literature has layers like geology, Egeria lived in the

Palaeozoic. She finds a hill too steep and the Euphrates too wide ('frightening'), records a gift of apples and addresses her recipients as 'Dearest ladies' or 'Loving ladies, light of my heart'. Who was Egeria? She was probably from modern Spain or France; she was educated and solvent; she may have been a nun. She wrote in Latin. Parts of her letter, which vanished for seven hundred years, resurfaced as the book *Peregrinatio*. In it Egeria writes of her visits to Job's tomb in Karnaia (now Al-Shaykh Saad, Syria), the well where Genesis describes Rebecca's encounter with Eliezer, and the Sea of Galilee. The book has a fireside tone antithetical to male-authored Classical texts – 'You know how inquisitive I am' – and it is personal: in Seleucia Egeria records her delight at meeting again a deaconess with whom she had made friends in Jerusalem. She changes plans often, fortunate to have handy mitigation, 'impelled' as she is at every turn 'by God'. She finds the tomb of Abgar 'old-fashioned'. Egeria wants to make the reader see what she saw. 'The mountain is on your right,' she explains, 'as you come from Egypt, and it was on the right for the children of Israel as they fled from Egypt and began to approach the sea.' Spontaneous use of yeasty direct speech makes the prose rise like dough. *Peregrinatio* has the advantage of the new, as a travel book: the author was reporting a breaking story. Fourth-century emperors had begun to raise monuments to the new faith – the Church of the Resurrection on the site they called Golgotha had gone up only thirty years before Egeria rode a donkey into Jerusalem. The First Council of Nicaea was almost within living memory. Scholars have scrutinised *Peregrinatio* to illuminate the evolution of Christian rites. The woman behind the veil goes on speaking after her death, like all writers.

Academics use the phrase 'religious tourism' to describe modern *peregrinatios*. Firms offer package tours to Bethlehem, Mecca, Masada, Fatima's shrine, the Kumbh Mela – sanctified places where the observer might deepen her own faith by absorbing the spirit of pious others. In cultures like my own, over which the grip of formal religion has loosened, religious tourism has evolved into *spiritual tourism*. What is a spiritual tourist? An individual who pursues truth among the shards of organised religion, more often than not, in the words of academic Fatima El-Tayeb, to leave behind secular European modernity and trigger a self-transformation in an exotically timeless Oriental landscape. Most women who take to the road with this noble goal seek victory over the self; subjugation of the ego and freedom from its bondage. It is odd therefore that the phrase 'to find oneself' is used to explain the motive of the truth-seeking middle-class when they decamp to Goa. 'To obliterate oneself' is a more accurate description of the spiritual tourist's goal. I can't imagine how a person could *not* wish to achieve victory over the self. It has to be the ultimate freedom. A person – me, let's say – can pursue the self-quashing goal just as well at home, but a loosening of daily ties and routines does, I find, facilitate proceedings. That said, other cultures embrace travel as a spiritual endeavour with authenticity. The indigenous people of Australia go walkabout 'when the heart is hot'. Perhaps it is the intention that counts, even in sodden Britain where nothing including the heart is ever hot.

History also reveals uxorial duty as an agent for women who travel, or should I say, as a travel agent. Those who write on the subject are often diplomatic spouses. I have observed the diplomat's life in many lands. It seems to me disagreeable, involving as it does the suppression of individual thought and the acquisition of standard-issue curtains which rise in quality as the diplomat ascends the foreign-office ladder. A hospitable British high commissioner once hosted a dinner for me at a Residence in East Africa. 'You'll see,' his wife told me as we came down for drinks. 'Alfred will open the gate at precisely seven-thirty, the guests' drivers will sweep them in, and at ten the same process will happen in reverse.' Events proceeded as foretold. Was this living, or revolving like figures on a Swiss clock? In Chile I visited some of the remaining political prisoners from the Pinochet era. It was quite an organisational business. After the event, the Second Secretary from the British Embassy telephoned, inviting me to lunch for a debriefing. 'What were they like?' he asked. I couldn't help wondering why he didn't answer his own question by visiting the *presos* himself.

Many women pack bags not to follow their husband but to get away from him. In 1739 Elizabeth Justice declared, in her book *Voyage to Russia*,

> The Occasion of my going into Russia was owing to my Husband, who was to have pay'd me an Annuity of Twenty Five Pounds a Year: Which he omitting to pay for Five Years, and a Quarter, I then, contrary to my own Inclination, was oblig'd, by my Sufferings for want of that Money, to go to Law; and I did obtain a Verdict in my Favour.

Justice secured a job as governess to the three daughters of an English merchant in Petersburg. There she saw the old-gold hues of Anna Ioannovna's Russia and heard thunderclaps louder – so she wrote home – than the ones in Hatton Garden. A judge at the Old Bailey meanwhile found the husband guilty of theft, sparing him transportation and sentencing him to exile for life (Henry Justice – nominative determinism was not at work in his case – chose Italy). He was 'obsessed with books', which makes one like him a little more, and books had precipitated his downfall after he stole some from the Library of Trinity College, Cambridge, his alma mater. Henry was Lord of the Manor of Rufforth, a Doomsday village near York, and after Cambridge had been called to the Bar. Elizabeth, née Surby, had grown up in London. She was the first woman to write in English about travels in Russia.

In a book I called *O My America!* after John Donne, I followed the trajectories of six middle-aged women writers who escaped that place called home. Why did they travel? Catherine Hubback was among Jane Austen's army of Hampshire nieces. Born in 1818, she married a barrister and had three sons before the husband entered an asylum from which he never emerged. Once the children were grown, at the age of fifty-two Hubback took off. Neither widow nor wife, she needed an identity and a modest income: these were her reasons. It was just after the golden spike had joined two bits of a railway line in Utah and made America one, shifting the frontier to San Francisco. Hubback rode the new rails to sunlit California. To get away from home is motive enough. 'Palled for the moment with civilisation and its surroundings,' Lady Florence Dixie writes in *Across Patagonia* (1880), 'I wanted to escape somewhere, where I might be removed from them as far as possible.' Many readers, she noted, would recognise that particular brand of dissatisfaction 'when one wearies of the shallow artificiality of modern existence'. This in a twenty-first-century context means abandoning the washing-up just as much as escaping the bondage of self; perhaps one wishes to leave the Fairy Liquid more urgently than anything. The self-escape goal is anyway impossible. You wander to the far side of the Taklimakan Desert (the name means you can get in but you can't get out) and find yourself sitting under a date palm looking glum. I know; and I expect you do too. Horace famously expressed it in an Epistle to a friend who was fond of taking trips: *Caelum non animum mutant qui trans mare currunt* ('They change their sky, not their soul, who rush across the sea'). 'Rhodes and fair Mytilene,' Horace presses on, 'are about as useful to a healthy man as an overcoat at midsummer or a loincloth in snowy winds.' Bestselling author Isabel Savory came to the same conclusion in 1900 in *A Sportswoman in India*:

Too often travelling is a Fool's Paradise. I am miserable; I want to get out of myself; I want to leave home. Travel! I pack up my trunks, say Farewell, I depart. I go to the very ends of the earth; and behold, my skeleton steps out of its cup-board and confronts me there.

In her book of Indian travels, Savory reveals another reason the nineteenth-century female ventured abroad.

It is unkindly said that the gentler sex are shipped across to the East, provided with costly trousseaux, for the mere purpose of meeting gallant captains and prosperous chief commissioners, noble Benedicts who for many years have run the gauntlet of the pick of the very limited ladies' society up country, coming unscathed out of the fire, and are only destined now to fall before the latest coiffure from home.

'I am afraid,' Savory continues, 'this old wives' tale no longer holds water.' Like every other form of travel, once everyone went out to India to husband-hunt, the thing was spoiled. Too many women flowed east and too many men came home on three-month furlough as ships sailed more frequently and the voyage shrank. 'In short,' Savory concludes, 'familiarity and close inspection betray the copper through the Sheffield plate.' But she reported that 'time has changed the Memsahib too', and many women had begun travelling to India and indeed to the whole world simply 'with the object of seeing other sides of that interesting individual, man'. Savory observes in India that 'Englishmen are supposed to possess an insatiable desire for slaying something; a healthy-minded woman has invariably a craving to do something.' British men perceived the Antarctic as another mammoth to be slain outside the cave.

During the pandemic, when I looked up guidelines for foreign trips, official rubric asked, *Reason for Travel*. Was it for studies, or business? For military deployment, or a certain category of medical care? At that baleful time none of my motives were eligible. They weren't important enough. Spiritual tourism was off the agenda, and so was abandoning the Fairy. It didn't matter why you wanted to go, as you almost certainly couldn't. I began to wonder how much the not-going mattered. How important had it all been, in fact? After half a lifetime on the road, I know, or I think I know, travel has enriched my experience. But did my grandmother, who was born in 1912 and never ventured beyond the sound of the village bells – what Italians call *campanilismo* – have a less rich emotional life? I think not.[1] Where, do you imagine, does this leave us? On the hard shoulder, or in the departure lounge?

The Antarctic was 'a country that made all the other countries seem strange'. The quotation is from a Maeve Brennan story. She was a writer who helped me understand my polar experiences. I travel with a dozen old postcards to set up around each temporary billet, whether room, igloo or tent. In one of the pieces Brennan wrote for the *New Yorker* she called herself a 'traveller in residence', and her biographer suggests that 'For her, life is a journey whose destination remains unknown, and perhaps the most satisfactory home is a boat, or a train . . . Maeve is a passenger, not a driver.' That was me. The cards I have carted around the world for decades depict familiar images such as Botticelli's perennial chart-topper 'The Birth of Venus'. Over Antarctic days and nights I thought about Venus's Medusa hair and regretful face and the bulrushes in the foreground, and the flowers falling around, and about what it meant to me, and the way in which perpetual light shone upon her through the ice bricks of an igloo geologists had built for fun. Venus's skin was even more luminous than Sandro intended. The postcards protected me in a place where no flowers grew; against what I don't know, except that it must have been my own demons, as there were no polar bears prowling the West Antarctic ice sheet.

Brennan put my travels in context. 'It was a lesson in brilliance,' she once wrote of a moonless night in East Hampton, 'and what

---

1. My paternal grandma lived to see my first book. I asked her once, after I had finished travelling but before I had begun to write, if she had any advice. 'Don't try to be clever or funny,' she said without hesitation. 'Just be yourself.'

the sky said was "Everything that happened happened a long time ago".' She herself lived not in igloos but in Manhattan hotels and apartments, or in upstate New York, sometimes behind the long Atlantic beach that runs from the Hamptons to Montauk. Domestic arrangements seemed always to be temporary: even her marriage was temporary. When she lived in Snedens Landing (now Palisades) in Rockland County, Betty Friedan was next door researching the book that became *The Feminine Mystique*. Brennan's personal malaise was not quite the one with no name that Friedan exposed (surely the same one Isabel Savory experienced when she longed to leave home), but she refers often, in her work, to Hans Christian Andersen's 'The Little Mermaid', the fairy tale which sets out a young girl's choice between staying where she is loved and safe, and losing everything in order to see and know a larger world. The postcards were the stanchions of a bridge between the worlds.

Brennan grew up in Ireland when houses were cold and food plain. Her parents were both nationalists, and when Éamon de Valera sent her father to DC as Fianna Fáil's Secretary of the Irish Legation, the seventeen-year-old Maeve also shifted across the Atlantic. (Bob Brennan later became the Minister.) She worked for seven years, starting in 1943, as a fashion copywriter for *Harper's Bazaar*, then moved to the *New Yorker*. During her long tenure at the latter she spent prodigiously at Saks Fifth Avenue and had the bills sent to the magazine, which paid them. Brennan was barely five feet tall and wore a bouffant that a colleague said was taller than she was. She was Irish but not really Irish American, even with JFK in the White House (they were the same age, Catholic, and built of County Wexford stock). Brennan wrote a lot about freaks. Her women in particular were often physically odd: large, ungainly and dour, like the Fat Slags. Their appearance symbolises a lack of belonging for which the glamorous Brennan's clothes and high heels never compensated, perhaps why she ended up a bag lady. As a writer she insists on truths that

cannot be reached by facts alone and uses external appearance
as a tool, which is what the best travel writers do. That is why
their stories transcend place to reveal human nature in the way
only great writers can.

Women's motives for travel are various, as I have tried to show.
When I went to the Antarctic in the early 1990s I had few female
companions on the page. I had to imagine how the Egerias and
Savorys and Justices would have tackled the ice, emotionally
and physically. I can't say I felt any sense of responsibility for
being the first to set my experiences on paper from there and in
that way, but does one feel much of a sense of responsibility for
anything at thirty? There were no female polar explorers. Three
schoolgirls wrote to Shackleton offering their services on his
1914 expedition (the letter survives in the Scott Polar archive in
Cambridge). *We*, they began,

> 'three sporty girls' have decided to write and beg of you to
> take us with you on your expedition to the South Pole. We are
> three strong, healthy girls and also gay and bright, and will-
> ing to undergo any hardships that you yourselves undergo. If
> our feminine garb is inconvenient, we should just love to don
> masculine attire. We ... do not see why men should have all
> the glory, and women none, especially when there are women
> just as brave and capable as there are men.

The Boss replied with 'regrets there are no vacancies for the
opposite sex'. The anecdote is generally retailed as a joke, but I
wonder if it was, to the three sporty girls. Progress came, slowly.
Some sources state that the first female to set foot on the seventh
continent was a 'waitress' on a Norwegian whaling vessel, but
I have never heard of a whaler running to waitresses. Others
say it was Caroline Mikkelsen, the Danish-Norwegian wife
of an explorer, in 1935, though she may have only stood on a

peri-Antarctic island. The wife of the leader of one of the first US expeditions to the far south, Edith 'Jackie' Ronne, née Maslin, accompanied her husband Finn, a former naval commander, as a working member of an expedition based on Stonington Island in the late forties. She may have been the first to step onshore. Jackie wrote press releases which went, eventually, to the equivalent of an agency pool (actually the North American Newspaper Alliance, one of the trip's sponsors). She had not intended to go south. She went to Beaumont, Texas, to wave the ship off. Norwegian-born Finn suggested she stick with him till the Panama Canal. There, he said, well, you might as well come on to Valparaiso. And so it went on. The only reason Jackie went in the end was because she had no reason not to. Perhaps that is the best reason of all. Seven members of the Ronne team objected in writing to Jackie's inclusion, on the grounds that it would 'jeopardise their physical condition and mental balance'. They relented when Finn Ronne announced that Jennie Darlington, the chief pilot's wife, would also go south. Apparently two were less jeopardising than one.

The fifteen-month Ronne mission disintegrated in a bitter storm of interpersonal conflict, as many expeditions do – the Antarctic seems to bring out the worst in men, testosterone-wise. In preparation for my own trip I read many accounts, and American Will Steger's *Crossing Antarctica* (1992) sticks in my mind in this regard. The author, in my opinion, takes rather too much trouble to snipe at Geoff Somers, the Englishman on the six-thousand-kilometre dog-sled traverse. The fourth person to reach both Poles, Steger, having completed the formidably challenging dog trek (one storm lasted over a month) and gained worldwide acclaim for the record he had set (fifty-two countries bought broadcast rights), wrote the inevitable book and was still concerned with putting the boot in to an apparently harmless teammate. And they had billed it as a multi-national expedition fostering cooperation in the polar regions. Like most men who appear on book jackets sporting a frozen beard, Steger is now a climate change activist. His website reveals, 'Will Steger established a Trust in 2013 to preserve Will Steger's legacy.'

Other men panic lest they might not appear the biggest guy. A picture researcher on a book about a two-man unsupported crossing of the Antarctic told me she was not permitted to select images depicting one of the sledgers in the rear. The British polar community still harp on about Vivian 'Bunny' Fuchs' superior role to that of Ed Hillary in the 1955–58 Commonwealth Trans-Antarctic Expedition, the first to reach the South Pole overland since Amundsen in 1911 and Scott in 1912. It was Sir Vivian, then director of the British Antarctic Survey, who said in 1964 that conditions on base were 'too arduous' for women. We had split the atom by then and the moon landing was only five years in the future.

Disputes notwithstanding, the Ronne expedition added to the sum of human knowledge. 'Perhaps most important,' *National Geographic* reported, 'the explorers at East Base had finally proved Antarctica was all one continent, laying to rest the theory that a frozen sea divided it.' Jackie Ronne died in 2009. And the first female at the Pole? In the 1969–70 season six American women researching in the Ross Sea area travelled to ninety south in a resupply aircraft and walked down the cargo ramp arm-in-arm.

At McMurdo Station, the largest Antarctic base, women worked in a variety of roles, from the galley to the heavy shop, and they were present at almost every science camp I visited. These included a two-person team lowering flow-capturing gear into fast-running ice streams at the bottom of the West Antarctic ice sheet and a two-hundred-strong group at South Pole Station sending balloons into the atmosphere and maintaining the most complex telescopes ever manufactured at temperatures of minus forty. Both Soviets and Americans had been quick to despatch women south once their polar programmes achieved lift-off (the British had not). Irene C. Peden was the first female scientist to work in the interior. She was an associate professor of engineering, conducting research into the lower ionosphere. It was 1970, and she was forty-five. The Navy laid the many obstacles on her path south, she later said, not the National Science Foundation, which administered polar programmes. The Navy operated logistics, and the admiral commanding the South Pacific fleet was the most vociferous opponent of women. He kept going on about bathrooms. Decades later Peden wrote, 'I was staggered to find that the first woman astronaut, Sally Ride, had to put up with the same stuff.' At McMurdo, where she transited on her way to Byrd sub-station, Peden noted, 'They had a nuclear power plant at McMurdo then too, and the pipes were lying round on the surface melting the snow.' Residual radioactive material was less important than women's lavs. It is hard to fathom the extent to which men have channelled their antipathy to women into toilets. In the Second World War 'the latrine business' dominated discussion on the admittance of female correspondents into the military press corps. Lieutenant Colonel Philip Astley, director of press relations at British HQ in Cairo in 1941, repeatedly briefed the War Office on the 'unacceptable inconvenience' of mixed facilities. When a female reporter visited one British battalion in the desert, Astley wrote, the problem was so acute that not a single soldier was able to open his bowels, to the agonising extent that 'at least three hundred men were unmoved for three days'. Mountaineer Arlene Blum cited, in her 1980 memoir, a letter she received from an expedition leader inviting her not to join a forthcoming climbing trip to Afghanistan. 'One woman and nine men would seem to me to be unpleasant high on the open ice, not only in excretory situations, but in the easy masculine companionship which is so vital a part of the joy of an expedition.' Yeah, just like Steger said when he spent three hundred pages laying into his teammate. Blum called her book *Annapurna: A Woman's Place*, but *Excretory Situations* might have sold more copies.

A female colleague accompanied Peden to her Antarctic research camp: again men cultivated some vague idea that safety lay in numbers and that danger might be averted if more than one alien joined the team. In an account of her experience from the point of view of female practicalities Peden addressed the critical bathroom issue: 'My husband Leo had bought me a port-a-potty before we left, and I took it along . . . I made everyone else get out of the van, and I stayed in it . . . Julia [the accompanying woman] on the other hand, made everyone stay in, and she brought the port-a-potty out.' I am not sure what to make of this Aesopian fable. But I admire Peden for getting on with the (non-potty) job. An official had told her that if her visit failed, female scientists would not return to the ice for many years.

I have a weakness for nuns and am pleased to usher in a Benedictine, Michigan-born biologist Sister Mary Cahoon. She was one of the first two women to overwinter in the Antarctic. The year was 1974, and the pair (yes indeed) were studying temperature adaptation in krill, other invertebrates and fish. They were able to do this based in McMurdo, since the station is on the coast, though of course Sister Mary and her colleague had to drill a hole in the frozen ocean before they could haul out crustaceans. Over the dark months, 129 people were in residence, the majority Navy personnel. All received Sister Mary with kindness. She enjoyed her experience, except for a period of bereavement after a young colleague died in an accident. They kept his body frozen (one of the few tasks in the polar regions which does not introduce a confounding host of difficulties) and shipped it home when the sun rose and a plane came. In the meantime, Sister Mary wrote, 'There was a service but no burial; there wasn't any resurrection joy.'

I did not have Sister Mary's certainty about much, if anything, when I watched a quaver of smoke at the top of Erebus or sat at my desk later while rain lashed the ill-fitting window. What did I believe? I called myself a feminist. As a child I never heard of feminism, or saw it in action, or learned about it from the hump-backed television permanently nictitating in the lounge. Women's issues did not feature as a topic of debate at my girls' secondary school, where I had a full scholarship on account of the direct grant system (the 1944 Education Act created direct grant grammars in England and Wales; fewer than two hundred existed when I was a pupil). Our teachers were mostly spinsters who had lost fiancés in the second war. The year I went up to Oxford – 1980 – marked the seventh female intake at my college in 471 years. In the first term my Greek tutor, Reverend Leslie Styler, born in 1908, told me often, in Homeric asides, that 'women aren't suited to epic poetry'. I must have absorbed feminism osmotically. I caught the last decade of second-wave *Spare Rib*, went to (non-university) consciousness-raising groups that gave currency to the second wave in a house in Oxford's Cowley Road where we were not unmoved in any sense, and read Simone de Beauvoir. Once I relocated to London I took Feminist Theory night classes in Camden Town at the Working Men's College (you can see why I don't write fiction. I don't have to make it up). There, as the 24 bus thundered past another ill-fitting window, bell hooks and her cohort began to inform my view of sporty girls in the world and on the road.

In addition, I was a member of the Labour Party. I have said that I come from a tribe of blue-collar Tories who hated everyone who wasn't like us. The *Daily Express*, ever-present in our flat, corroborated family beliefs in the failings of the poor to lift themselves up and of the disenfranchised to obtain the necessary totems of inclusion or leave for another land (and never come back). It wasn't for me.

Third, I was a Christian of sorts, though not by upbringing. My maternal grandparents were practising Methodists, as I indicated, but God didn't trouble my parents. I was on the lookout,

and after standard teenage experimentation settled into a branch
of high Anglicanism perched on which you can believe anything
you want or not believe in anything much. I could dimly see, and
I see more clearly now after decades sitting in on other cultures,
that a desire to reach for the transcendental is a fundamental
part of being human. Perhaps it is the part that separates us from
emperor penguins.

Faith and feminism intersected in 1992. I was in my bedroom
in Mornington Crescent, the north London street where I lived for
many years, arranging laundry on the radiator having returned
from a newspaper assignment in Nepal the day before. Bizarre Inc
was at number one in the charts with 'I'm Gonna Get You', and
the song was playing on the radio. A flash news announcement
crashed the 5 p.m. pips. The General Synod of the Church of
England had passed a vote sanctioning the ordination of women.
Damp high-altitude thermals in my hands, I looked out of the
window at blue sky. So we were gonna get them; we had in fact,
got them. I had the same feeling twenty-two years later when
the Synod sanctioned female bishops. Will.i.am had just moved
to the number one spot with 'It's My Birthday' (it had been, in
March – his and mine). I remember not crashed pips then but
my mother's irritation. I was with her in her Bristol flat when we
heard the news. 'Women,' she said, crunching salt and vinegar
crisps, 'have to get in on *everything*. First rugby. Now bishops.'
And that was that, bar my ongoing quest to find a rugby-playing
female bishop, perhaps to jump out of mother's birthday cake.

Were these semi-beliefs linked to motive? I wasn't looking for
anything particular on my travels; I wanted only to see how other
people did things, and, in the case of the Antarctic, how the world
looked without human beings in it. And to find out whether anything was true. I wore the Phrygian freedom cap but it seems to
me now that I was immature in my objectives (as if it weren't the
function of old age to take that view). I don't mean that I should
have been Che Guevara. But it would have been productive to

be a little like him. I intuitively tried to see things through other people's eyes – what would be the point of any other kind of looking? My goal was mutual understanding; you can say that's pious, and of course I had to pay the gas bills too, but so does everyone. I did not then realise that the prejudice my generation inherited would not wither and die, but transform, in the hands of populist demigods, into a new weaponised racism, and that the backlash would outlaw the quaint practice of looking through others' eyes.

My own environmental awareness dawned during the Antarctic phase of my life. Thirty years had elapsed since Rachel Carson wrote *Silent Spring*, and in the lower latitudes scientists were aware of climate perils ahead. The emerging emergency had not yet entered the public consciousness to a meaningful degree, but glaciologists and other climate scientists had fought for research funding which might wake the world. I sat with Hermann Engelhardt on the ice eating from a tube of Pringles next to what looked like a bent and windblown tee flag in a golf hole. Drillers had bored through the West Antarctic ice sheet to reach an ice stream a kilometre down. They already knew that the ice streams, wide as the Amazon, move at hundreds of metres per year, whereas the unstable sheet pressing down on them shifts a few tens of metres. (Like icing flowing off a wedding cake, ice on the surface of Antarctica is slowly but persistently rolling towards the coast, forcing its way between and over mountains, turning itself into glaciers split by crevasses and spilling out into a floating ice shelf – either that, or at the end collapsing into the Southern Ocean. I say 'slowly', but it rolls now more quickly than it did when I was there.) Engelhardt and his Caltech team had sent down equipment to monitor basal slide and, they hoped, learn more about ice drainage dynamics. I wrote in my notebook, as we sat there, 'Hermann understands the basic physical mechanism that controls rapid motion and thinks it may have consequences for worldwide climate change. How does it affect sea levels if ice is discharging quicker than it used to?' Data was the revolver on the table in Act One of a Chekhov play. The mechanics of ice streaming did turn out to play a role in the response of the ice sheet to climatic change, and now, a generation on, every school-girl can tell you a story of apocalyptic inundation, every economy in the developed world has committed to carbon reduction, and even the *Daily Mail* accepts that Engelhardt's hunch was correct.

My creased photographs of that day reveal the pair of us smiling in sunshine under the flag, the red wing tip of a Twin Otter poking into the frame. It was a good time. I actually wasn't locked out, and as Woolf continued her essay, 'I thought how it is worse perhaps to be locked in.' Hermann returned each summer to Upstream B, transiting through McMurdo where he picked up a Hercules which landed on blue ice at a seismology camp. From there he and an assistant went on in an Otter to Ice Stream B to haul out their equipment and collect data. They usually had to replace the flag, which had lived through its own private fight with the katabatic winds of winter and deserved peace. In the photo, Hermann is dipping his hand into the Pringles tube. The potato-based snacks were so ubiquitous in American polar research camps in the early nineties that I wondered why the manufacturer didn't use images in its ads of rosy-cheeked scientists ingesting them in the lee of a winsome glacier. Back at McMurdo the innocent product pack-aging met another budding branch of environmental awareness: recycling. Now, with consciousnesses raised all round, the activity has entered the ordinary kitchen the world over, but then it was new, and the National Science Foundation was determined to lead the way in the division of waste products into recyclable categories. I remember standing in a shed on base after a stint in one camp with a dozen empty Pringles tubes in my arms and having a minor nervous breakdown in front of an array of bins labelled: thin cardboard, thick cardboard, foil, hard plastic, soft metal, aluminium – and so it went on, *with the dismantled Pringles tubes claiming a component in each category.*

Another photograph from that season shows a VXE-6 helicopter swinging an external payload on a rope over Lake Fryxell. The cargo looks like a yellow crate with a gleam in the paint finish (cadmium?) but is in fact a block of frozen urine. Waste disposal in the mostly snow-free Dry Valleys of Victoria Land where it has not rained for two million years is subject to ferocious controls, as nitrate chemists there work in the most pristine conditions on the planet. Choppers fly out all human waste. VXE-6, the airborne squadron of the US Navy, did all the flying then for the American programme. They called themselves the Ice Pirates. I called them Testosterone Airways.

The idiosyncrasies of Pringles tubes notwithstanding, I can see, looking back, that it is only because experts took recycling seriously then that we take it seriously now. NSF decision-makers educated us. Similarly, in response to more readily apparent risks, they heavily promoted safety training, calling it Survival Training to invest it with a sense of urgency. I went on many courses for which I was and remain grateful. And the polar traveller is not only walking on thin ice. He also has to face psychological anxieties of absence. When I was at South Pole Station, residents tittered grimly about the toll separation takes on relationships. Someone had pinned on the galley noticeboard a quotation from a letter a physicist allegedly received from his girlfriend after nine months away: 'Yours is bigger, but his is here.' In winter, the vagaries of mental health pose a particular hazard. Inuit have regionally varied words and phrases for the particular brand of lassitude, anxiety and depression associated with the dark months. In the seventies an overwinterer at a Soviet Antarctic base became involved in an argument with a co-worker over a game of chess, seized an ice axe and murdered the other man.

As I said, it is a serious business. But let me get this out of the way. To stop it happening again, the Soviets banned chess.

*

Like a business conference, the Antarctic facilitates romance, bathroom seals notwithstanding (Tom Stoppard once said that the hotel room constitutes a separate moral universe; imagine how much more separate, when there are no telephones, no roads and no time zones). One was unmoored. The unmooring was part of the attraction: like many, I went to the Antarctic in part because I loved the idea of putting the trappings of civilisation aside. Also like many before me, I felt the spiritual effect of the singing ice when there is no gas bill dropping through the letter-box, and no letterbox. At my first field camp, in the Dry Valleys in Victoria Land, the twenty-four-hour light cycle released us even from the strictures of diurnal routine. Except of course it didn't in a practical sense: the five-person team was just that, a team, and they had to cooperate. The PI (Principal Investigator) used to stand outside the Jamesway (communal cooking and eating tent inherited from the Korean War) and bang a saucepan with a mallet to wake the grad students. It might be three in the afternoon, but another twenty-hour shift had to start, and anyway it was time to haul some fabulously expensive piece of nitrochemical-measuring kit out of the water that lay below a three-metre lid of ice. Because it never got dark there were no shadows. Everything was in plain view. Writers in particular had nowhere to hide. And I was the only writer.

Everyone who has read my book knows that I enjoyed a romance in the Antarctic. I left out almost all the detail, but the little I put in exposes me to ridicule now. Well, I exposed myself. When Francis Spufford edited his two-volume *The Ends of the Earth: An Anthology of the Finest Writing on the Arctic and Antarctic* in 2007, he chose an extract of mine about Seismic Man, the romancee. (Spufford had to include something from *Terra Incognita* as he was so short of Antarctic material by women.) I met the research engineering associate from Texas Geosciences on my first day at McMurdo. Personnel had not yet assigned me a room (not in this case because they were reluctant to do so), and as a result I was clumping round base in the standard-issue blue moonboots, giddy with a crazed sense of relief that I had finally got there. He too was waiting to go into the field, though he was not giddy, as he had done it all before. His role was to bury nitroglycerine in the ice sheet and set off explosions which monitoring systems on the surface recorded, contributing data to enable other people to map Antarctic land below its coating of ice. I gave him the moniker Seismic Man. The book did well in the US after an unexpected rave review by chief *Times* critic and Pulitzer-winner Michiko Kakutani. Nobody ever found out why she had reviewed that particular book, insignificant in publishing terms. That may really have been luck. But I discovered the power of *The New York Times*. Anyway word had got out in the small polar community and for years afterwards the director of my engineer's institute in Austin, when showing guests round, would fling open the victim's office door and say, 'And this' – pointing – 'is Seismic Man!' Readers with a good memory apparently exclaimed, citing a line from *Terra Incognita*, 'Ah! The man with the come-to-bed eyes and been-to-bed clothes!' It must have been annoying.

Saucepans and mallets aside, living in the Antarctic interior for many months was a love affair in itself. I meant it when I wrote then that the landscapes the most inimical to life, like the boyfriends, are the most seductive. We stake out our own banana skins and ring them about with converging arrows boldly labelled, 'Step here'. I can't say I regret incorporating material about Seismic Man in *Terra Incognita*, ridicule notwithstanding, but I am mighty glad I did not include more. It's private. In the decades after Antarctica I have met Seismic Man once, when he appeared in London in transit from some remote field camp. He was still setting off explosions; still the same in every way. I remember our time together fondly (perhaps not the frostnip episode). But my two months with the artist float into my mind with almost painful longing. We set up camp on 21 August, the day after the sun peeped above the horizon. It was minus 22, and the Transantarctic Mountains glimmered in a violet haze I had not seen before. The scientists had not yet arrived for the season, and nor had the Adélie penguins (only male emperors sit out the winter). In an hour the sky turned from wholly dark to wholly light, and each day we had more light: everything changed quickly, like a speeded-up natural history film of a flower opening. We had used a tracked vehicle to tow a couple of fish biologists' huts out to the sea ice and sit in them for two months to do whatever it was we needed to do before the arrival of piscatorial scientists. As members of the Writers & Artists programme we had W numbers, just as the scientists had S numbers. Lucia, the painter, was W-001 and I was W-002. We were the Woos, and our camp appeared on maps (sea-ice maps are redrawn each season as contours shift) as Wooville or, sometimes, as The Ant Art Chicks. The Fat Slags died there under the *aurora australis*. Or was it just pretty to think so?

2

Moni Sou?

Greece

A piece by Lebanese author Hanan al-Shaykh struck me as I struggled to become a writer. In the early nineties al-Shaykh had travelled to Cairo with a colleague to advise on a television documentary about Arab women. In the opening two pages of her essay, 'Cairo is a Grey Jungle', she confronts the dangers of generalisation:

> I was asked for the truth about the Arab woman and the repression she suffered ... About the sword poised permanently over her neck. As if millions of Arab women were frozen into a single figure with no past or future and all Arab countries have had the same customs and were narrow-minded or open in equal measure.

The proposed documentary sought to reveal the figure of the 'miallima': an Egyptian woman of authority with 'a penetrating gaze exuding severity and also desire, plump under the gallabiya which is designed to hold her curves but in fact accentuates them'. (Its current English translation might be 'girlboss'.) She is 'a liberated woman in a position to tell others what to do'. In her essay, al-Shaykh, who was born in 1945, walks down

Al-Sukariyya and climbs to a mosque roof, where she observes 'The still waters of the Nile, on whose surface a person could have slept.' At El-Ghouri bazaar she comes upon a miallima eating a sandwich in a shop selling Ramadan lanterns in the shape of Mickey Mouse. Al-Shaykh asks the woman, who is wearing a black jersey, long skirt and 'modern-type makeup', if the shop belongs to her. 'Of course,' she replies. 'Otherwise I wouldn't be standing here eating like this. I'd have sneaked off into a corner where no one could see me.'

Going anywhere, to write about anyone, the writer risks over-simplification of the kind al-Shaykh identifies in 'Cairo is a Grey Jungle'. Generalisation lies like a baited trap. I saw it lying ahead of me before I even began my first travel book. What about individual agency? The weight and variety of experience? Thought? And what about the amorphous throng of 'Greeks' at the heart of the book I planned?

Before tackling the generalisation theme, al-Shaykh introduces another characteristic of life on the road. 'I am confused, torn between my love for all I see,' she writes, 'even the crumbling ruins and the delicate stonework almost buried in the dust of the past, and a yearning for the comfort of the hotel which is like an extension of the European country where I live.' I remember disembarking from the last mailship on the second leg of the St Helena–Cape Town run – a week-long journey across the South Atlantic – and finding myself on the tenth floor of a five-star hotel in a room that was all glass. It overlooked ocean you could sleep on, and you could not see Robben Island. I was in tenth-floor heaven, and guilt didn't set in till morning. Guilt at not pounding the streets to gather _material_ so that I could _write_ better and have more _insight_. One longs for the duvet sometimes. Or to stand in the corner eating a sandwich. 'I felt weak, and so homesick,' Julie Donnelly wrote from the slopes of Himalayan peak Kala Patthar in _The Windhorse_ (1986). 'I would give anything right now to be back in my house where I can run around all the rooms easily without any help.' Donnelly is blind. Her achievements as a female traveller tower over those of us who continually fail to see the wood for the trees, or get out of the wood altogether before the trees arrive. I did not though share al-Shaykh's dilemma, 'Where do I belong? In the East or the West?' When I approached my first book I can't remember feeling conflicted about anything. Despite a shortage of role models in life or on the page, it did not occur to me that my status as a woman ruled me out of anything. I did not want a career as a miallima, telling people what to do. I knew I was a writer. I can't imagine what gave me the confidence to test that conviction. I don't remember feeling anything approaching confidence. Naiveté is the gift of youth. Maturity hands over the imposter syndrome.

Why Greece, for my first book? It would not have occurred to me either to start writing about any another place. I had studied Ancient and Modern Greek at university, and to learn to speak the modern variety I had lived in Athens for a year in the early eighties, working for a small literary publishing house. The Greeks were at the halfway point on the arc of their leap from donkey to Mercedes: in the country, where I travelled at weekends and during holidays over the course of that year, if I asked how far somewhere was, a person sometimes replied, 'three cigarettes' (meaning how long it took to smoke them, not the distance laid end to end). In villages, they would demand, when I tottered in alone having walked from some whole-carton spot, 'Are you Christian, or foreign?' – eisai christiani, i xeni? There is not a person alive in Greece today who would understand that question. But I often heard it. The identification between being Greek and being Christian had roots in the Tourkokratia, four long centuries of Turkish occupation when Greeks were Christians and Turks the enemy. Turkish overlords actually turned the Greek Orthodox Church into the engine of the country's administrative infrastructure. Church was state, in a real sense, albeit within a hostile superstate. Many older Greeks (generalisation alert) perceived themselves as Christian rather than Orthodox, since as far as they were concerned no difference existed between the two. This link between past and present (Are you Christian, or foreign?), manifest so overtly in Greece, led me to my first book, and a preoccupation with the theme.

I was twenty-one when I moved to Athens for the publishing job. My work involved collecting book orders and riding the tram to downtown bookshops from the firm's three-room office in Mets, an old district in the south of the city. The office was at the top of two steep flights of outdoor steps, and the tram stop was at the bottom, opposite the Temple of Olympian Zeus, scavenging ground for hundreds of cats. The benign owner of the firm, who lived on an island, was seldom in Athens, and the only other

employee was a bookkeeper who worked one day a fortnight. So I was almost always on my own, and, as in all small businesses, did everything, including, in winter, feeding the wood-burning stove with discarded wooden shoe-lasts purchased as fuel. I carried books in a canvas bag slung over my shoulder and delivered them with invoices I had cut after customers rang in orders. Some titles were in English – our bestseller was a history of *rembetika* music – and all concerned Greek culture. On my rounds I often stopped for a thimble of gritty coffee with the friendly booksellers at Folia tou bibliou, Eleftheroudakis and a series of tiny shops that were little more than dusty cubbyholes. Some of the younger staff became friends. Everything was cheap. We went out dancing so often that I kept a frock at the office, as sometimes it was eight in the morning before we called it a day. (Songs I remember dancing to every night in strobey Athenian clubs include 'Gloria', 'You Can't Hurry Love', 'It's Raining Men' and 'Eye of the Tiger'.) At that time the wholesale meat market on Athinas Street kept a series of small restaurants that fed meatpackers and nightclubbers, and we would sit alongside one another, blood-soaked overalls next to glitter dungarees. This was before the EU took its cleaver to the Dimotiki Agora and insisted on refrigeration and other nonsense.

I lived not far from the office, high up in Pangrati, minutes east of the marble Panathenaiko Olympic stadium. My studio flat was on the ground floor of a *polikatikia*, a modern, six-storey block with a corner shop attached. I used to meet the neighbourhood matrons in the *pantopoleion* ('everything shop') in the mornings, all of us in nighties, purchasing a bag of coffee or a tin of honey. It was from them I learned how to make staple Greek dishes that I still turn out regularly. ('Do I squeeze a lemon into the tzatziki?' SNORTS: 'Do you want it *PIKRI* [bitter] *to tzatziki sou*?') The window of my main room looked onto Kiniskas Street, and naturally, when I was there, the blind was up and the window open. A passer-by regularly poked his or her head in to ask, 'How much do you pay for this?'

Snow fell that winter, and from the communal roof I could look over at the Acropolis glimmering on its fresh white carpet. The speckled Parnitha mountains arced behind, the lower sentinels of Agra and Ardettos standing watch on either side. To me Athens was a beautiful city, more beautiful than Paris had been when I had lived there three years previously. It was less sanctified, and I somehow had a stake in it, which I had never had in Paris. In Athens I began to understand that one can belong anywhere. In *My Ántonia*, her superlative Great Plains novel, Red Cloud-raised Willa Cather writes of the joy of being 'dissolved into something great and complete'. That was what belonging meant: the place did not have to be literally great like a Nevada plain or an Antarctic snowfield.

From March my friends and I went to the islands every weekend. Hydra, Spetses, Tinos, occasionally further afield on a plane up to Limnos or over to Samos. Nobody had any money to speak of – I was virtually on a student budget – yet everyone had a modest job and we could afford anything we wanted, even taxis and the occasional flight on Olympic. We shared clothes. Staying with my employer in the country I worked the grape harvest and the olive harvest and enjoyed the great *glenti* that was Easter. I had a boyfriend from a communist bookshop who took me to Athenian beaches in the evenings on a scooter. At home we carried wicker-clad bottles down to Megarites' taverna where our friends the waiters refilled them with retsina from a barrel. The place in Platia Varnava, which everyone called Costas's but which officially had no name, had no menu either. Each night Costas himself reeled off the three or four dishes. In these small tavernas we listened to proper music – the Greek kind. Theodorakis, of course. *O psilos* (the tall one), as people in Mets and Pangrati referred to him, was part of the fabric of life, mentioned as if he were a friend. I never glimpsed him, or heard him play, an omission I regret. Bouzouki was at the root of his music, despite the fact that he trained as a classical composer. He had not yet joined

Mitsotakis's (centre) right New Democracy government, or written operas. All that was to come. The colonels' dictatorship had ended eight years before I moved to Athens. It was still fresh in the collective consciousness, looming as the Second World War had in my childhood. Theodorakis and his music represented something essentially of the Greek people; his songs made his working-class audience proud, and allowed them to feel superior to the filthy elite who had supported the dictators, passively, or even actively. And they were superior.

Friends and others piled into the studio on Kiniskas, often swinging their legs over the sill and entering through the window. Like most modern flats in Athens the separate kitchen had a window onto the *photagogo*, the light shaft at the centre of the block, and it was through that aperture, my neighbours told me, that cockroaches entered. Once, returning from the bathroom in the night, I saw a roach scuttle under the mattress. I didn't have an actual bed, and nor did I have a full-length mirror, and as we were always swapping clothes, this latter was a problem solved by opening the front door of my flat and entering a small lift opposite which did have a mirror. This regularly resulted in the lift ascending containing a preening youth who grinned goofily when a startled neighbour opened the door somewhere above. Once, I went up in a bikini.

About halfway through the fourteen-month stint in Athens I began dreaming in Greek. This too was part of belonging. In some ineffable way I now shared something of my friends' inner lives as well as their external counterparts. Much later, I came to associate this with glimpsing – as I imagined – the inner life of the dead writers I worshipped and who often seemed as real to me as people with whom I actually did live (as well as more agreeable). I went on to learn a few other languages, but I never dreamed in one in any meaningful sense. I had a nightmare in Russian once, but that was due to the tortuous complexity of verbs of motion.

A few years later I wrote my first book. I had the idea in 1989, the year the Berlin Wall fell, and signed the contract in 1990, the year Mrs Thatcher fell. It was to be the story of a six-month journey around Evia. The geographical homogeneity of an island appealed to me, as did Evia's little-known status abroad despite being the second largest Greek island after Crete. Greek books sometimes refer to it as the third, because they count Cyprus, and I liked that too. Situated northeast of Athens, Evia – the ancient Euboea – is almost part of the mainland, as a bridge joins the two at Chalcis, where the Evian Gulf narrows to fifty metres. On Evia the sun rises over the Aegean and sets over the Gulf, and at one or two high places in the middle you can see both coasts. Looking through my yellowing snaps I see again Sarakiniko in the north-east, where everyone in the village had the same surname and rounds of *mizithra* goat's cheese dangled from rafters. I see the Dragon House on Ochi, the second highest mountain, probably a temple erected by migrant mine workers from Asia Minor during or after the fourth century BCE. Mythological dragon monsters had assisted by hauling the enormous stones up the steep slopes. There is Vangelitsa, the goatherd's daughter, dancing at her wedding. I stayed with her parents on her wedding night.

I can't remember what time we went to bed but I do remember getting up a few hours later and realising I had to catch a ferry shortly. The Greek insistence on hospitality can kill you. As I sat at the table feeling as though I might have died, Vangelitsa's mother set before me a jug of retsina mixed with Coca-Cola and a plate of cold goat.

At Kavodoro in the southeast, the most isolated region, I remember a couple at Vrestides winnowing wheat on a circular threshing floor. I stood on my own not far from there on the spot known in antiquity as Cape Kafireas where King Nauplius, son of Poseidon, held torches high to trick Greek triremes rowing home from Troy. He was mad with grief for his son Palamedes, killed by his own countrymen at Troy, and stood on the Evian cape luring Agamemnon and his fleet onto the rocks. The sailors thought Nauplius's lights were guiding them to shelter. Euripides wrote about it in *Helen*. A mist rose off the water as I stood there looking down. Any sailor would have followed a light. Meanwhile at Amigdalia people arrived with plastic containers strapped to donkeys and mules, looped with brightly woven *tagaria* (square shoulder bags) bearing the pattern of their village. A hose had piped water from the mountains since the wells ran dry three years previously. The first language of the community was *Arvanitika*, an Albanian-based dialect. (Multitudinous strains come under its blanket heading in deep-rooted 'Albanian' communities round the world.) The Kavodoro Evians had spoken their version since the fifteenth century without any contact with what we call Albania. Their ancestors arrived in 1402 when the ruling Venetians needed, for defence purposes, to replenish a dwindling population.

One night in Kavodoro I planned to sleep on the beach. A yacht had anchored in the bay. Not a superyacht, but a luxury one. As I was unrolling my sleeping bag at dusk, a couple appeared on the sundeck and sat opposite one another at a table. The vessel was so close, and the night so clear, that I could see they were wearing evening dress. White-uniformed crew glided round serving dinner. The still-warm air was like a balm. Then, as the rim of the sun slipped over the dark horizon, a rotund figure in white tie and what looked like tails emerged onto the top deck. The moment lay in suspended silence, the water of the bay reflecting only a gibbous moon. The fat man raised an arm, and '*E lucevan le stelle*' from *Tosca* rang between the Kavodoro cliffs as the stars did indeed shine. '*E non ho amato mai tanto la vita*' ('And never has life been so dear to me').

King Nauplius lowered his flaming torch.

Men and women from Ancient Evia sailed to what was to

become Italy and founded colonies like Cumae near modern
Naples. After the triumphs and catastrophes of the Classical and
Roman eras, and the long course of Byzantium, Venetians and
Lombards set up in feudal residence for 265 years. The island's
strategic position on a sea route between Europe and Asia meant
that Evia marked the eastern frontier of the Venetian empire
and when, in the fifteenth century, La Serenissima lost control
of the island they called Negroponte, the doge and his dukes
and freemen wept bitter tears. The past beat inside Evia like a
second heart. The central slice was rich in thirteenth-century
Byzantine churches predating even seagoing Venetians. Some
were attached to monasteries (the Greek word *monastiri* includes
nunneries), all open to guests in the Greek way. I often stayed
with sisters high up behind Vathia or the black ridge of Mount
Kandili. They said their *apodipno*, the last service of the day,
in the narthex of their chapel. A single oil lamp cast flickering
light over a biblical cosmogony while black shapes crouched
on the chequered floor. A sketch of this scene appears in my
notebook, always a sign of heightened emotion as I am a poor
draftswoman. At the Last Judgement, men and women climbed
a ladder to heaven, angels hovering above willing them up and
little brown beasts below dragging them into dragons' jaws. In
one chapel the beasts were pulling two men off the ladder by
their beards. Unfriendly Turks had gouged out the eyes of many
figures in the frescoes using tools which had left marks like
snowfall. The nights I spent high on an Evian mountain were
among the most meaningful of my life. But what did they mean?
Allied to the impulses associated with reading and travelling,
I was increasingly interested in the spiritual landscape, inner
and outer. You can call it religion, but I was already thinking of
it as a primitive grope towards the transcendental – the characteristic I mentioned that distinguishes us from penguins. My
involvement with Greece stimulated deeply felt passions I did
not yet understand.

Greece was now in the EU, which was very much outside the transcendental realm. The country had joined in 1981, but the euro was still a long way off when I was in Evia, and Yanis Varoufakis had only recently started shaving. Greece was derogating itself from supranational legislation and public finances were nowhere near as parlous as they would become. In 2015 Jean-Claude Juncker, then head of the European Commission, was rewriting history when he said, no doubt with Monty Python in mind, that Greece joined the Union because 'We [Brussels] didn't want to see Plato play in the second division.' Greece had applied for entry in 1975, but over the following years the Commission noted that the country's inclusion would 'pose serious problems for both Greece and the community'. France was worried about the impact on farmers, Germany about cheap migrant labour, and everyone feared involvement in Greece's dispute with Turkey. According to the *Guardian*, when the true cost of the debt emerged, 'European grandees say it had been "a mistake" to allow Greece to join the currency.' But Plato, a robust centre-half, played on in the top tier.

'Even in Greece,' V. S. Naipaul wrote in *An Area of Darkness*, 'I had felt Europe falling away.' When I was travelling around Evia with a carpetbag the country didn't seem particularly European. Even then western commentators and sages who revered Ancient Greece joked that its modern equivalent was an economic basket case. The paradox fermented tension that still exists, in Greece itself and in the western narrative of the country. This was not a path I followed. I set out not to discover if Greece were European, but to find an island culture. It quickly emerged, as I got on and off buses, that Evia doesn't have one, internal geography, notably mountains, having fostered a series of isolated regions. (The peaks had forestalled construction of an airport, which in turn had thwarted the development of package-tour resorts.) I soon began to make out something more important. Women in Kavodoro wore shoes made out of tyres and nowhere did Evia resemble Athens. But below the surface in both places a new world order was fundamentally altering the way people perceived themselves. The journey I made in the eighties was as much about time as place. Perhaps all journeys are, but this one seemed especially temporal, then, and certainly now, and it set a certain tone for the decades to come. The fundamental shift I noted was one that moved away from any awareness of the spiritual dimension. Hellenism, a series of ideas or ideals wrapped around Greekness, has persisted since Ancient Athenians argued in the *stoa*, or so some believe, this author included. I witnessed it weakening. Atomisation had reached remote villages on lonely capes. Greece was becoming more like everywhere else – or was I *generalising*?

On the ground rather than in the clouds I didn't fit in to anyone's scheme of things, Hellenistic or not, because I was on my own. *Moni sou?* was a phrase of which I tired. I heard it whenever I pitched up in some village or hamlet. 'On your own?' Solitary status seemed weird, apparently to everyone. There was nothing Greek about the attitude. When I started travelling for my second book, I heard only *Sola?*, the South American equivalent of *Moni sou?* I could tell you that phrase in many languages, as it has pursued me around the world. But I am only really travelling when I am alone. I am anonymous. Paul Theroux once wrote, when his companions left, 'I was glad – it meant I was alone, I was safe, now no one would ever know me.' Solitary status therefore offers protection and solace, not the opposite. It appeals to the travel writer in particular, because it represents what he or she is trying to uncover. G. K. Chesterton said, 'If we wish to depict what a man really is, we must depict a man alone in a desert or on a dark sea sand. So long as he is a single figure he means all that humanity means; so long as he is solitary, he means human society ... Add another figure and the picture is less human, not more so.' You only have to look at a Hopper painting. Protection and solace yes, but a whiff of ethical ambiguity clings to the writer's self-righteous solitary status. 'How easy it was to lie to strangers,' writes Chimamanda Adichie. 'To create with strangers the versions of lives we imagined.'

The Greeks nonetheless treated me like a freak because I was alone. Perhaps they were the freaks. Who decides who are the freaks? I am not about to compare myself to a war correspondent. It must be obvious to the reader by now that I have none of the skills required. But I fall with interest upon stories about the female reporter's isolation in the field, the *Moni sou?* attitude. Clare Hollingworth broke the story that Germany was about to invade Poland after she observed tanks, armoured cars and field guns mustering in the valley below the frontier road from Hindenburg to Gleiwitz. The next day she arrived in Krzemieniec, where the Polish government and foreign embassies, having fled Warsaw, had set up on the hill. The British ambassador was already settled into a hotel with his team. 'No family?' he asked Hollingworth when she said she was with the *Daily Telegraph*.

I was in my late twenties when I stumped round Evia, yet was called a *koritsi* or *kopella* – a girl. In all languages, the word for *girl* carries less respect than the word for a married woman of the same age or younger, perhaps much younger. The role I was perceived to play could not carry a serious content, as far as the onlooker was concerned. I was figuring all this out, or starting to. At the same time I had to learn that it was all right to be miserable on the road, as I sometimes was, and to feel that travel was a waste of time, as in reality it never was. Here's Martha Gellhorn, another war correspondent:

We can't all be Marco Polo or Freya Stark but millions of us are travellers nevertheless. The great travellers, living and dead, are in a class by themselves, unequalled professionals. We are amateurs and though we too have our moments of glory we also tire, our spirits sag, we have our moments of rancour.

Gellhorn goes on to explain 'rancour' by asking who among us has not felt, thought or said, in the course of a journey, 'Why do they have to make so damn much noise?' She writes of feeling 'frayed and bitter', a peerless phrase which I considered for the title of this volume. This was al-Shaykh's guilty 'longing for the comfort of the hotel', a longing exacerbated, as both she and Gellhorn recognised, by the spotlight that shines without respite on a lone female. Rural Greeks in particular were keen to belabour my status as spinster abroad. They expressed anxiety over my inability to find a husband, often wondering if it weren't to do with my *fakes*. Freckles must be uncommon in Greece, as many interlocutors pointed at my inherited abundance while frowning and keening, and some suggested the application of lemon juice to fade them out, thereby improving my looks, and, it was hoped, marriageability.

A final point on the 'girl' word. Let us free it from its chains! It's not acceptable for my father's colleagues to call office assistants in the building firm 'girls' (or to pinch their bums and display nude calendars, as dad did his whole working life). But I can, if I want, rehabilitate the word to celebrate how I feel when I am with girlfriends, or when I wish to express solidarity – in the book on America I mentioned I refer occasionally to my six middle-aged subjects as 'girls'. I enjoy the intimacy the word confers and I like the way it makes men feel excluded. When language distorts a woman's place in the world, I want it to evolve. Honorifics similarly invite gender-based prejudice. I'm talking of course about the tiresome 'Mrs or Miss' issue. Decades of foaming resentment have resulted in online purchases arriving at my house addressed to 'The Right Reverend Sara Wheeler'. A small protest in the drop-down options box. 'Mrs' and 'Miss', however, were not conceived as indicators of marital status. 'Mrs' began its career as an abbreviation of 'Mistress', a title which in early modern England was the female equivalent of 'Master'. Both designated a person with social or commercial capital who directed servants or apprentices or taught pupils. In other words, 'Mrs' accorded the bearer respect without marital connotation, as 'Mr' did, and does. Take Mrs Eleanor Coade, who in 1766 bought a ceramics factory on the south bank of the Thames and perfected the clay-based material now called coade stone. You can still see it ornamenting Brighton's Royal Pavilion. The capable Eleanor was known far and wide as Mrs Coade not because she was married (she wasn't), but because that was the title of eighteenth-century businesswomen. Professional respect extended to the stages of London's West End close to Mrs Coade's factory. Scholars have recently tried to identify an enigmatic thespian by the name of Mrs Jones. The poet John Keats several times referred to her in his correspondence. Keats and Mrs Jones had a thing going on. He saw her first in Hastings; she played an Aeolian harp; they exchanged grouse and other game birds. Sleuthing academics pursuing the woman, having failed to find a married actress matching Keats' clues in the records, have recently concluded that the divine Mrs Jones did not have a husband. In modern parlance, she was Miss Jones.

I listen for ways in which other tongues deal with the Mrs–Miss dilemma. In Siberia, everyone in the Republic of Sakha called me *edjii*, a term of respect in Turkic Sakha and one accorded to older women irrespective of marital status. It's like 'elder sister'. A younger woman in the Republic is *kyys*. Notwithstanding the delicate business of whether the addressee has indeed transitioned from nubile lovely to old bag, the Sakha way does not seek to define a woman in relation to a man. I quizzed a Sakha friend about this. 'It was even easier,' she said, 'under the Soviets.' Moscow insisted on the use of Russian in its vigorous attempt to suppress the myriad indigenous languages of the Union. 'Then,' my friend said, 'everyone was "comrade" – *tovarisch* – which can be used for men and women.' This 'comrade' wheeze is useful, but the introduction of communism seems a heavy-handed solution to the collateral problem of packages arriving in north London addressed to Squadron Leader Sara Wheeler. (That was a phase, between 'Right Reverend' and 'Lady'.)

The woeful 'Ms', reminiscent of the sound a bee might emit in a children's story, is not even new. An anonymous American first proposed its use in 1901. In the seventies, US lawyer, House Representative and all-round heroine Bella S. Abzug proposed that the 'sex prefix' of Mrs or Miss be both eliminated in Federal records and correspondence in favour of Ms on the grounds that the two terms are discriminatory and invasive. Half a century later, in the English parliament we have a Mr Speaker or a Madame Speaker. Why? Because everyone knows the honorific 'Mrs' fails to confer adequate respect. 'Mrs President' has not yet been put to the test, but the term has a dying fall. Marital status is an obsolete piece of information when it comes to filling out forms. The consistent deployment of Mrs and Miss sends the message that women should be defined in the context of their relationship to men. I remember, as a little girl, absorbing the information about female titling. Transitioning to Mrs, it seemed, was a settled part of the natural order. You needed a man to make the grade. A stigma attached itself to women of my mother's age who had not passed the social test; it was an exam Miss Fox, our kindly, grey-haired lollipop *warden* (best I can do), had for mysterious reasons failed.

What kind of message is that?

I had some modest experience of travel and life in strange lands before I embarked on the pursuit seriously for my first book. In 1981 I volunteered for two months on a kibbutz near Ashkelon, just north of the Gaza Strip. I picked pears with a machine like one of those in which a person goes up and down to change streetlights, and walked behind a faulty combine harvester lifting potatoes a hopper had spat out in error. The envelope-making machine in the kibbutz's factory was also faulty, and I often spent the night shift snoozing over the conveyor belt until the thing jolted back to life and wrenched me from a dream in which I was trapped behind the transparent window of a glued-down A4 envelope, screaming to get out but unable to make any sound. I admired the kibbutzniks hugely for confronting every obstacle with dignity (and I liked them), especially as there were a lot of obstacles. Looking back, I admire them even more. Only fourteen years before my visit, the President of Iraq had declared, 'Our goal is clear – to wipe Israel off the map', and when Cairo Radio ('The Voice of the Arabs') followed up with appeals to *Slaughter, Slaughter, Slaughter*, crowds took up the cry in the streets of the capital. I wonder if Yoni, the man tasked with repairs to the infernal envelope-maker, thought of those events in Cairo, and many others of the same kind, as he bent over piston rings and connecting rods. One did not speak of such things.

I mentioned Paris. I lived there for a year before university, working for Wallis, a British high-street fashion chain trying its hand among *Parisiennes*. It came about like this. Since the age of fifteen I had worked as a Saturday girl at the Wallis branch in Bristol's Broadmead, and usually during the summer sale and in Christmas holidays as well. With that behind me, I managed – I can't remember how – to persuade the firm to deploy me in its newly opened shop in the City of Light. I travelled from Bristol to Paris on the night ferry, via London, Dover and Dunkirk. My parents, who waved me off at Bristol bus station, were familiar with the night ferry as they had seen it featured in an episode of *Steptoe and Son*. It turned out to be the end of the line for the night ferry, as it closed later that same year.

I had limited experience of France. My school operated an exchange programme with a lycée in Bordeaux, the port with which Bristol is twinned. The year before O-levels a cohort of us were parcelled out to families in the arrivals hall of the still single-terminal Bordeaux airport. I went to a château. It was a wine château – Beychevelle, in the Saint-Julien appellation on the left bank of the Gironde estuary, just north of the confluence of the Garonne and the Dordogne. My exchange partner's father was not the owner but the manager, and sometimes at night when frost set off alarms at the end of the *vignobles* the phone in the hall clamoured and Monsieur Ruelle got up to go and heat the vines (I never found out how). The Ruelles, who occupied a modest wing of the château, were a hospitable family. I enjoyed eating new things at their table. They dropped prunes in empty coffee cups and scooped them out with a teaspoon, and served radishes on their own with butter. When Chantal and I took the school bus to the lycée in Pauillac, Madame Ruelle packed us off with Nutella baguettes for lunch, an indulgence beyond the dreams of Epicurus.

Before that exotic school exchange, when my brother's behaviour had not yet deteriorated to a point that prohibited holidays, my parents had twice driven us to Brittany in a camper van, taking the ferry from Portsmouth to Roscoff. We stayed on a

Eurocamp site and drank not Coke but Orangina from bulbous bottles. But like most ordinary English tourists to Brittany in the early to mid-seventies, we transported food with us, to avoid the expense and incomprehensibility of the French super-market and, I suspect, the perils of foreign comestibles. Garlic in particular was a known hazard of cross-channel travel. I remember driving down the *route nationale* from Roscoff behind a Cortina loaded with boxes of cornflakes, tins of baked beans and packets of Smash that filled the rear window. Food was an important component of travel then, much more than it is now or ever will be again. Because one ate different things Abroad, meals were indivisibly associated with the Other. We in the West Country knew little of foreign grub. The only takeaways in Bristol were Chinese, so deliciously loaded with monosodium glutamate that they gave you lockjaw, and pasta came in a tin. So did imported fruit. A half-consumed tin of mandarin segments resided permanently in the fridge. I realise only now that it was not always the same tin. Where did the canned mandarins go? It didn't mean our food was bad. My family ate vegetables from my maternal grandfather's garden in St George and eggs laid by Henrietta, a hen alive for so long I now see they must have concealed a series of gallinacean deaths to spare my childish feelings and deployed 'Henrietta' as a generic hen name. We picked plums at Bisley and berries elsewhere. Intensive meat production was in its infancy. Working-class and lower-middle-class families ate plenty of processed food in the 1960s, though not as much as we consume today. But imported products? No. The avocado, like the high-speed train, had not yet pulled into Temple Meads. There was little disposable income for fancy food anyway. There was also little if any foreign travel in the course of which one might encounter new dishes. As I have said, families like ours did venture to western France, but I did not know anyone who travelled long-haul. Package tours began to filter onto the pages of the *Daily Express* in the late sixties. Spain captured the market. We never went, but our neighbours did. They brought back a bullfighting poster with my name inserted as star *torera*, and a lurid liqueur that crystallised around the neck while languishing at the back of the lounge cupboard. We made do with *Carry on Abroad*, which appeared in 1972. At the Wundatours half-built resort on the Costa Bomm sand came out of the bath taps, the local tipple Santa Cecelia's Elixir gave the drinker X-ray vision and Lily and Marge's wardrobe did not have a back so the man in the room opposite could see in. Do I need to say that the meals in the film were inedible? If we went to a foreign land, though, we had to eat foreign food, as we could not take enough Smash to last a week. Distrust ruled the roost on which the *poulets* slept before we bought them from a spit revolving behind glass in Prisunic. As the decades wore on, my biological tribe went abroad with increasing frequency while the tribe I was making for myself more often than not hailed from that once forbidding and forbidden territory. All sorts of food integrated itself into the British diet, although 'integration' was the key. My mother still pushes a Tesco croissant down into a toaster every morning before larding it with margarine and marmalade. And if only we welcomed refugees as warmly as the chicken tikka pizza.

We did not look out onto the world in my childhood. Did the world barge in anyway? Not much. The only global event I recall in secondary school was the 1973 oil crisis, which meant fuel was in such short supply that Redland High closed and parents organised a rota, individuals I recognised ringing the doorbell with Cyclostyled sheets of work. The pre-photocopying system cranked out paper printed in purple ink, which smelled agreeably of what I now know to be methanol. (Everyone smoked in the staff room, so I don't know how we didn't all go up in flames.) The three-day week that continued into the following year resulted in lay-offs both at my father's building firm and at the Wills tobacco factories. But the shadow of war did not lie over us, as it had lain over the childhood of both my parents. I was a boomer.

Paris at eighteen came as a shock. I wasn't ready for it, despite the fact that I had already lived alone for a year after the disintegration of my parents' marriage. It took me time to settle in. But I did in the end, and was sorry to leave, a pattern that was to repeat itself. A French teacher at school had hooked me up with a young professional couple who had a flat in a not particularly smart part of the 16th *arrondissement*, and I stayed there for the first nights until two friends of theirs said I could rent their *chambre de bonne* round the corner on rue la Fontaine. I liked the idea of inhabiting a Fontainian fable, perhaps *La Grenouille Qui Se Veut Faire Aussi Grosse Que Le Bœuf*, in which a frog tries so hard to grow as big as an ox that she explodes. My landlady and landlord were single professionals who went to an office every day, but I worked shifts, as the shop was open till eight, and also on Saturdays, for which I had a *jour de congé* midweek. The woman of the flat, a likeable, brusque secretary at the OECD, was a keen cook, and when she bought rabbit at the La Fontaine butcher (there was also a *boucherie chevaline*), the man who shared with us put the needle of the blocky stereo down on 'Bright Eyes', a hit the previous year.

As for Wallis, the shop was a disaster. Hardly any customers even came in. It was on an underground floor of the Forum des Halles, a new centre in the 4th *arrondissement* on the site of the old fruit and vegetable market next to the Beaubourg, also known as the Pompidou Centre, which Jacques Chirac had opened three months before I arrived. Its brightly coloured external heating ducts (it was one of the first inside-out buildings) were still news. 'An ocean liner run aground in the Marais,' locals had said after the press conference at which the winner of Pompidou's design competition was announced in 1969. Some were still saying it when I arrived and it was not a scale model but the real thing. The Marais, where the fruit and veg market had operated, had been part of Paris for many centuries (the name means the Swamp;

the fertile Marais grew most of the medieval capital's produce). As for the Beaubourg, of which I grew fond, I see it now as both a late flowering of 1960s utopianism and a foreshadowing of the showy architecture that would define urban regeneration in decades to follow. Next door, in the Forum des Halles, Wallis was adjacent to FNAC, which should have been a good location. The days were long. Madame Régnier, the shop manager and a former Dior model, rendered us half-mad by insisting on playing Pink Floyd's brand new *The Wall* on the shop sound system, as I recall it permanently. I still hate the songs, even though they make me think fondly of the *girls*. To my great relief and good luck the other shop assistants, mostly in their early twenties, were fabulously friendly. There was Maria, who lived nearby with her Sicilian parents; Claudine, whose handsome boyfriend called in every day; and Pharinette, a clever, beautiful young woman of Cambodian heritage. I don't know why she worked there, as she was rich, finishing a master's, I think, and preparing to get married to a *bon chic, bon genre* Frenchman about to graduate from the *grande école* ENA. There was plenty of time to talk as we stood around for hours with nothing to do but tidy the rails of white trench coats and jumbo-cord trousers with braces. The girls invited me to many places, from nightclubs to Sunday barbecues in the *banlieues*. We patronised the cinema, though not the antiseptic precursor of the multiplex that had opened in Les Halles. The girls preferred dives in the 14th. The new wave had crashed onto the beach by then, but we did see Truffaut's final major film, *The Last Metro*, the week it came out.

Most of my colleagues had grown up outside Paris, and when we could coordinate our rare Saturdays off several invited me to their family homes for the weekend. There in Sedan and Charleville-Mézières I met parents and grandparents for whom memories of the Second World War were fresh. Every single one of them, it turned out, had worked for the Resistance. It is amazing how many men and women in France had striven to uphold the noble goals of *liberté, egalité* and *fraternité* in the Vichy years. The girls themselves never talked about politics. Neither the progress of the Fifth Republic nor that of the world beyond interested them. There was certainly no awareness of the interconnectedness of past and present I mentioned earlier. The Forum was a commercial failure (they should have stuck to fruit and veg) but no sense wafted between the clothes rails that this mirrored any general decline, and that recession was waiting in the changing rooms. We did not connect individual events to broader patterns. We were focused on our own lives. I do though remember that various fathers and brothers in the industrialised north of France lost jobs.

On my day off I would walk the streets, nosing into small, independent shops, of which there were plenty in the 16th in those days. I can't remember any chains at all, except Félix Potin grocery stores. Giscard d'Estaing was in the Elysée. He had recently radically reshaped the public service broadcast landscape, and five years previously had mandated the creation of Radio France. As its headquarters, Maison de la Radio, was at the end of our street due to the proximity of the Eiffel Tower transmitters, we observed comings and goings. The Round House, as the most famous landmark in that corner of the *seizième* was known, broadcast and recorded classical music as well as news and drama, so players hurried past our front door dressed in black with humped cases in their hands or on their backs.

Yakking all day with the girls complemented the French I had learned from schoolteachers who hadn't spent much time on the other side of *La Manche*. I started reading poetry for leisure that year, I think partly because immersion in French stimulated an interest in the elasticity of language. And in its quirks. Why in the shop did we call the colour I knew as burgundy *bordeaux*? Why did my friend call her parents' French windows *fenêtres anglaises*? Why was *crème anglaise* never served in Bristol if it was so English? Similarly, customers' gabbled questions often communicated their meaning before I understood the actual words – a process intrinsic to poetry. Of course, in Wallis communication still regularly foundered on an aspirated subjunctive or a customer's *accent du Midi*. Nobody seemed to mind. Or perhaps memory is working in my favour.

Looking through my photographs, loose now in the album as the fixative has dried out, another thought shoulders in after more than forty years. It concerns the men and women I had met at the girls' houses who said they had worked for the Resistance. I believed them. Then as the years went by I ceased to believe them. But now, thinking of al-Shaykh's warning against generalisation, I suspend judgement. I had been looking at a photograph of me and Martine, an endlessly cheerful colleague at Wallis. Unlike the others, she was married, with two small boys. Once she had lent me her fondue set, carting it in from Ivry-sur-Seine on the RER. The photograph shows the pair of us in Ivry one Sunday, mugging for the camera in front of an austere brick building, weak sunlight filtering through the branches of a lime tree. I had written the caption 'Ivry?' I realise only now that tunnels ran under Ivry, where the Nazis starved *résistants* to death. The victims scratched farewell notes on the brick walls. I didn't know that at the time, and now I wondered, looking at Martine's smiling face, if the tunnels actually lay under those four feet. Who cared if a few people reinvented their past? It was the real past that counted, the true past scratched on a wall. Like memory, the past is a warpy reservoir. When I was living in Paris, the Second World War might, in my mind, have taken place at the same time as its Trojan counterpart which undid Nauplius. But as I said in the Introduction, it was recent: I can remember events that took place thirty-five years ago plain as crotchets on a stave. So could people I met in Paris in 1980. In the especially hard winter of 1944–5 the residents of 64 rue la Fontaine had fuel to heat the rooms for two hours a day, on a good day. In the fort at Romainville, not far away, the Nazis had built metal-lined

ovens, not yet to kill but to roast Resistance prisoners slowly prior
to interrogation. The past another country? I think not.

During the long Evian journey ten years later, which turned out
to be the first of many journeys, notions of identity crystallised.
'Coagulated' might be a better metaphor. Mrs Thatcher had been
prime minister since I was eighteen. I had internalised the notion
that in order to succeed one had to become an honorary man. My
first literature teachers at school (women) and university (men)
had inculcated the same idea. At school Miss Meade, an English
teacher who wore a silver bun, had set a term of Gerard Manley
Hopkins – a brilliantly bold choice, I now realise, for unworldly
young girls. ('Women have had less intellectual freedom,'
Virginia Woolf wrote in *A Room of One's Own*, 'than the sons of
an Athenian slave.') It was 1973, the year the lights literally went
out and there was no heating fuel, but the government had not yet
closed or partially closed schools, so we sat with our bottle-green
blazers wrapped round us and woollen tights as well as socks.
Miss Meade asked a question one day about Hopkins' thirty-five-
stanza ode 'The Wreck of the Deutschland' (*My heart, but you were
dovewinged, I can tell,/Carrier-witted, I am bold to boast,/To flash
from the flame to the flame then, tower from the grace to the grace*).
Susan Reid put up her hand and began her answer, 'I feel . . .'
Miss Meade leapt in. 'Think!' she said. 'Always think, not feel.' I
did not write 'feel' to describe a personal response to literature
for many years, but by the time I reached my late twenties I was
discovering that Miss Meade, the sainted teacher who pushed
open the door to books, the greatest gift of my life, was wrong. It
was all right to feel. That was in part the purpose of books. One
had to learn to harness thought and feeling. The personal was
political – keeping down girls' natural tendency to respond with
feelings belonged to a social structure that perpetuated female
inferiority. *You are a freak for being on your own.*

As for Woolf, she turned away from Athenian slaves and struck a more positive note. 'These evils,' she said of her peers' lack of intellectual freedom, 'are in the way to be bettered.' She then expressed the hope that,

> By hook or by crook . . . you will possess yourselves of money enough to travel and to idle, to contemplate the future or the past of the world, to dream over books and loiter at street corners and let the line of thought dip deep in the stream . . . If you would please me – and there are thousands like me – you would write books of travel and adventure . . . For books have a way of influencing each other.

Two, perhaps three generations on, Woolf's thoughts on the relationship between social structures and the inner lives of individual women rang clear to teens in Westbury Park, like a bell summoning us out and on. The words she used about the inimical soil in which many potential women writers grow – 'alien and critical' – were oddly familiar, Hopkins notwithstanding. But golly Woolf made you feel you could fight your way free of it all and find 'freedom and peace', and a room with a lock.

Tackling my first book, I looked at how other women had responded to Greece and let the line of thought dip into the stream. Few if any who had left written records had been alone. In most cases, the ticket abroad was part of a husband package (al-Shaykh's Egyptian women, the sword permanently poised, could only dream of tickets). One, who arrived in 1802, was Mary Bruce, Countess of Elgin, first wife of the seventh Earl. Since the second half of the eighteenth century, after Napoleon's armies blocked other routes south, tourists had begun to arrive in Greece on the Grand Tour. But the Elgins weren't tourists. Lord Thomas served as Ambassador Extraordinaire to the Ottoman Empire, and it turns out that he was not the only Elgin looting artefacts. In a letter to her husband in May, after he had left Greece, Lady Mary wrote from Athens, 'I have got another large case packed up this day – a long piece of the Basso Relievo from the Temple of Minerva.' Continuing to list her trophies, she ended the paragraph, 'I feel proud, Elgin!' In the next instalment of the letter, added the next morning, she enumerated eight 'cases' en route to England packed with antiquities. 'Do you love me better for it, Elgin?' We don't know whether he did or didn't, but he divorced her a few years later.

Americans began to ship in, among the first New York-born Julia Ward Howe (1819–1910), abolitionist, suffragist and author – she wrote 'The Battle Hymn of the Republic'. When she boarded a steamer in Argos heading round to Athens it was carrying, in full view, the decapitated head of a robber-brigand, one Kitzos. The head had a price on it. 'He had fallen with one eye shut and one open,' Howe reported, 'and in this form of feature his dissevered head remained ... with its long elf-locks lying loose about it.'

At about this time I began to read more women. Only then did I become a reader. On Evia, books and travel became the foundation for everything else, one looking inward, the other out. I began also to make connections between the two. That was the key. As Woolf wrote in *Three Guineas*, a much later work than *A Room of One's Own*, you needed freedom of mind and a critical independence. Hope glimmered on the horizon. The Berlin Wall had been rubble for a whole year when I started writing about Evia on an Amstrad, and the Cold War was surely over. There was room for optimism. How the fugitive years rob us even of past hopes.

I learned from Rebecca West about connections between inner and outer, history and lived life. She fused past and present – the link that had led me to write about Greece – more credibly, and more effectively, than (almost) anyone. In *Black Lamb and Grey Falcon*, the central book of her life, she describes a meal in a Mostar hotel. When people came in and hung up their fezzes and went to their seats and played draughts and drank black coffee, they were 'no longer Moslems, merely men'. West goes on, 'Young officers moved rhythmically through the beams of white light that poured down upon the acid green of the billiard tables, and the billiard balls gave out their sound of stoical shock. There was immanent the Balkan feeling of a shiftless yet just doom.' When I was in Mostar I remembered West's 'shiftless doom' and felt it all around.

Born Cicily Fairfield in 1892, West took the name of the heroine of Ibsen's *Rosmersholm*, a free-thinking, complicated radical. She wrote in many forms, like Gellhorn – reportage, fiction, travel. She was a feminist. In her travel writing she showed that men might well have all the power and women might well do all the work, but, as a result, women get the dignity in the end. An uplifting thought, my friends! Though you and the women do have to get through to the end. *Black Lamb* tells the story of three journeys in what was then Yugoslavia in the thirties (mostly with a husband, so West dodged ambassadorial insults). Published in two volumes in 1941, it is, like many of the best travel books, a kind of love affair: a deep engagement with a land and its people, as well as a threnody for a lost peace that may never have existed. West showed me that women writers can maintain a robust point of view without the loss of either credibility or style. Indeed the robust PoV is another key in the bunch, an inextricable component of *voice*, which at the end that is all a writer has – without it every door stays locked. I did not side with the Serbs as West did in the Balkans. But I saw the web of the past in which people of those lands were enmeshed as if the intervening decades had never poured out blood. Nobody in Mostar belonged in al-Shaykh's East or in her West. They belonged there. But there was no there there.

I returned to Greece half a dozen times over the ensuing decades, spending a month, for example, on Lesbos of migrant-camp infamy. I had written in 1990 that I saw in Evia an island 'on the cusp of rebarbative change'. Villages have gone on dying since then, and the threads of continuity and tradition I saw growing frail have weakened further. Or is it I who have changed, as one must, as life hurtles forwards across fresh borders? In the introduction to the second edition of *Evia*, which came out in 2007, I wrote this:

When I reread the book to prepare for this new edition, I could barely recognise the young woman who was me. Now, looking back from a mature (and more jaded) vantage point, I envy her those months of indolence and freedom when she simply leapt aboard whatever bus turned up.

My children were five and ten. Now they are twenty-five and twenty, and the random bus is again tooting, do I still struggle to make that young woman out? I continued that introductory paragraph, 'Regrets flutter down like homing pigeons: I wish I had carried on researching the Early Bronze Age; I wish I could still read Homer in the original; I wish I still went out dancing till dawn with strange men.'

3

Ladyland

India

I signed up next to write a book about Indian trains. My long-standing agent came up with a working title which duly appeared in the contract: *Hindu Choo Choo*. I loved Indian trains and I loved India. I spent a lot of time there in my early thirties. I had a cultural column for the *Calcutta* (as the city then was) *Telegraph* – still a newish venture – and once unsettled my colleagues at the paper's Chowringhee offices by arriving illegally, overland from Bangladesh. A former member of the Bangladesh national football squad had organised the bribe, accompanying me to the immigration building in Khulna wearing not the green-and-red strip but a three-piece suit. When I walked in at Chowringhee my colleagues said they would have thought I needed a cricketer to bat for me against Bangladeshi functionaries, not a soccer player, as nobody cared about that sport in Bengal.

I was at that time a correspondent for the American trade magazine *Publishers Weekly*, covering events and trends in Europe and anywhere else. I had just filed from the Delhi Book Fair for them, and for three other trade magazines in Britain, two of the pieces appearing under a different byline to fool innocent readers, as if anyone cared. This was a typical stunt of mine, and one that continued for decades: deciding where I wanted to go based on where writing was taking me (or I thought it was – often I hared off in the wrong direction), then harvesting commissions to fund the enterprise. On that occasion I had set off by train for Rajasthan as soon as the book fair ended, having eaten and drunk to agreeable excess at publishers' parties in the five-star-hotels of Delhi, which were more lavish than anything I had experienced on the literary scene in London. At any rate I was faxing in copy about book deals, new imprints and company mergers as I went round Rajasthan. Bullets weren't whizzing past my head or anything. But the backstreets of Jaisalmer were tough hunting grounds for typewriter ribbons. And there was no dancing till dawn with strange men, more's the pity.

I was often back in the east, and in Calcutta, where in clear dawn light citizens wrestled for recreation on the banks of the Hooghly. It was cool in April – by mid-morning the wrestlers would be queuing outside banks holding umbrellas. A communist government had gained control of the state legislature the previous year with a commitment to inward investment, but there was little evidence of action. The once-grand Great Eastern, among the most famous hotels in the world, continually popped up on *Telegraph* news pages ('feeds' were still comestible). A French chain had tried to buy the 1840s behemoth from the government, but union demands, allegedly including the maintenance on the payroll of battalions of deadbeat staff, had stymied the deal. 'So it's still state-run and crap,' said a friend. (The hotel entered private hands a decade later.) The daytime population of Calcutta was eighteen million then, its night-time equivalent thirteen million. Five million came in and out every day on a train, many not sitting in a coupe as I did but hanging from the outside.

After watching the wrestling one day, a colleague took me to the Shishu Bhavan children's home off Circular Road run by Mother Teresa and three hundred Missionaries of Charity sisters. Candy-striped curtains blew in the wind off the entrance hall and

nuns in white linen with blue borders hurried by, a silver cross
hanging at the top of each starched sari. A squalling hum rose in
volume as we walked up to the first floor where a blackboard read
BABY WARD male 68 female 98. Inside, under-ones lay in green
metal cots in issue clothing that included, for girls, red romper
pants. Balloons floated over some cots, anchored by string, and
piles of check cotton nappies lay on the floor; sisters changed
the babies twelve times a day and eight washerwomen laundered
nappies by hand. On the ground floor a similar ward housed toddlers. A sign there read, WE CANNOT SEE GOD LIVING WITH
US, BUT HE IS HERE. Rooms off the lobby were adoption offices,
and anxious white faces moved above low screens like moons.

   Mother Teresa, called just Mother in Calcutta, was popular with the middle classes in the north of the city, her Nobel
laureate status reflected in the pages of the *Telegraph*. Being in
the Shishu Bhavan made me see Christopher Hitchens' critical
book on the living saint, *The Missionary Position*, in a new light.
Hitchens argued that Mother cared more about Roman Catholic
dogma than about the poor, and that she lacked empathy. Later
he added to his thesis by saying Mother Teresa 'was not a friend
of the poor. She was a friend of poverty'. I have spent too long in
Latin America to maintain an overall positive opinion of the role
of the Catholic Church among the poor, but how could the care
given in those Calcutta wards be anything but good? I mentioned
this to my companion. 'There is black and white in everyone,' he
said, not moving the debate forward. He seemed to know what he
was doing though, and we walked two blocks along the street to
the eighty-seven-year-old Mother's house for her morning levee.
In the lobby a sister was dressing a wound on an infant's leg while
two other children stood alongside, and nearby a woman with
three engorged plastic bags petitioned another sister. We walked
through a courtyard and upstairs to an encircling gallery. Outside
one of the rooms stood the Mother. She was blessing a baby who
was leaving for her adoptive home, and seemed to be speaking

English. She looked about four foot. My companion had pushed me into the short queue. My turn came. Mother Teresa took my hands and said, 'God bless you.' I kissed her hands, she smiled benignly – there were about ten people in the small area – then she slipped behind a curtain, and a sister said softly, 'The Mother has gone for today.' Later that year, she went for ever.[1]

A third of a page in the *Telegraph* domestic news section ran stories concerning 'the northeast', the part of India that blooms from the stem of West Bengal's Siliguri corridor. Its five states and two union territories were apparently permanently bristling with insurrection, as ULFA (the United Liberation Front of Assam) kept the Indian Army on the run.[2] If ever I mentioned the word 'India' my colleagues at the paper chorused, 'there is no such thing as one India', and although they loved talking about their country and contradicted themselves constantly about the one-India question (it was only I who was not to mention it), it was the northeast that they painted in the most distinctive palette: a sensational land in the shadow of the Himalaya populated by peoples whose roots sank deep into a variegated history and culture.

For reasons that will become clear in the next chapter, I never wrote about my travels in the northeast. When I dusted off my notebooks for this volume I was astonished to find how much time I had spent in Assam and Meghalaya, and what I had forgotten. My first trip began at ten at night on 2 April 1997 when I boarded the Saraigat Express at Calcutta's Howrah for the twenty-two-hour journey to Guwahati, capital of Assam. I had not sat down on my blue leathereen berth in the second-class sleeper before a fellow passenger introduced himself as a railway

---

1. I stand by my observation that Mother Teresa's work did much good in what was then Calcutta. But I agree with Hitchens' assessment of her willingness to support right-wing regimes. Her complicity added up to more than tactical compromise.
2. Sikkim has subsequently joined the officially designated North-Eastern Region as its eighth state.

doctor, one of a two-thousand-strong team caring for 1.4 million employees 'the length and breadth of India.' It is pleasant to have a medic in the compartment.

3 April. Conductor starts whacking my seat at 6.40 a.m. asking for his 20 rupees reservation fee. After that, everyone goes back to sleep until a man got on selling cordless telephones.

Later in the day, as the train threaded through the thin gap connecting the northeastern states with the rest of India, I lay on my bunk listening to the doctor and another man, both in white pyjamas, chattering contentedly, like doves. At Guwahati station, where I alighted, at least a thousand people were thrusting up one narrow staircase. Mr Das found me, though, before I reached the top. He was a stringer for the *Telegraph*, tasked now to act as my guide in the northeast. I was very pleased to see him. We took an auto rickshaw to the RajMahal as buses and taxis were not operating on account of a four-day private transport strike. Billboards warned in Assamese of the perils of AIDS, the script of which makes barbed wire look poetic; I had copied some lines into my notebook. Armed guards stood outside my room at the Raj all night. The hotel had four stars but no hot water. The dining room was dark, with low chairs and half-life-sized carved wooden horses. Breakfast consisted of buttermilk *idlis* and tomato sandwiches with the crusts cut off. When a waiter drew the curtains closed, it was impossible to see the coffee cups on the striped tablecloths.

Mr Das and I became friends. He was a short, likeable individual who ricocheted between high excitement and gloom. He seldom missed an opportunity to pontificate on the Bangladeshi problem. 'Other northeastern states,' he said one day, 'have laws protecting tribals, mandating that non-tribals can't acquire land. There's some,' he said, dropping an octave as he spotted a group of Bangladeshis on the other side of the road, men in lungis and women wearing their saris draped over their heads. 'But here in Assam we have no such laws and therefore we are more attractive to Bangladeshi migrants. One year recently three million came into Assam.' The figure cannot have been right, but Mr Das didn't mind. (He pronounced the tea-and-oil state *Asham* rather than *Assam*.) 'In ten years Bangladeshis will outnumber us,' he predicted. They did not. The Assamese population totalled 36 million in 2021 of whom 14 million were Muslim, not just Bangladeshi. (Cordless phones, on the other hand, had proliferated.) Mr Das was right, however, when he said politicians had a vested interest in preserving the status quo because Bangladeshis were 'a vote bank – Indian citizenship is tacitly offered in return for votes. It is old-fashioned patronage.' Impotence in the face of the Muslim vote was a popular theme in the northeast narrative. Deve Gowda, Mr Das said, referring to the prime minister unexpectedly elected head of the ruling coalition in Delhi the previous year, 'his hands are tied'.

In Shillong, capital of the contiguous state of Meghalaya, I stayed at the Alpine Continental, the name referring to a wooden architectural style once endemic and now subsumed in concrete, but still my notebook records, 'Shill = 100 x nicer than Guw.' At the Bara (big) Bazar tamarinds rose in globular peaks or lay unshelled in husky brown pods. We ate pineapple slices sprinkled with salt and chilli powder, purchased from a stall, and looked down the market's narrow lanes. 'The King of Meghalaya,' said Mr Das with a sweep of his arm, 'gets a cut of it all.' He was in a permanent state of fury about 'tribals'' exemption from taxation and the royal family's receipt of an unofficial tithe in its place. The monarch, elected for a five-year term from a pool of royal blood families, was a Khasi doctor. A million Austroasiatic-speaking Khasi live in the northeast. 'Tribals hate Indians,' said Mr Das, explaining that the former resent the dilution of their culture by 'people of the plains', notably Marwaris who 'mop up' on business. Everyone mentioned federal neglect. 'Terrorist groups,' someone unidentified in my notebook said, 'have operated out of frustration – you can understand it. A seventeen-kilometre corridor divides us from the mainland [note the word] and a myopic government in Delhi can't see beyond it.'

Mr Das lived to highlight lurking dangers, but as we settled into the car one day he remembered that his role was to promote my enjoyment and comfort. 'Only danger for Indians,' he said, his face like a crumpled paper bag. On the road up from Guwahati, military police had stopped us at three different checkpoints, frisking the driver and Mr D and searching the glove compartment. They left me and my luggage alone. They had once caught Mr Das without his ID and taken him to a military camp where they beat him up. Not all 'tribals' hated 'Indians', though. Mr Das's own brother had married a Khasi from Meghalaya, and as Khasis are matrilineal, the groom 'had to go and live at her place in the hills'. At first the Das parents wouldn't have anything to do with their daughter-in-law and her family. But relations had thawed.

Meghalaya, 'abode of the clouds', had achieved statehood in 1971 when it seceded from Assam. A peak we visited outside town was formerly the home of presiding deity U Shillong ('Where does he live now?' 'He's moved on. Too crowded.') The spot overlooked pine forest as well as the whole town and lake, but 'we saw bugger all bec clouds came right in'. It was the abode of the clouds still, even if the god had moved on.

At the start of a seven-hour drive from Shillong to Assam's Kaziranga National Park, touted as 'a diversity hotspot' in the sub-Himalayan belt, Mr Das rolled down the window to throw coins at a Ganesh shrine. A group of barefoot women were passing with sheaves of logs on their backs. The hills were barren. Population pressure had reduced the old eight-year slash-and-burn cycle to two. As usual, government incentives to prevent the practice were failing. As usual, people were too poor to stop. Blackboards outside betting shacks offered odds on the daily archery at Shillong polo ground.

Halfway to Kaziranga, at Nongpoh, police vehicles crawled over the trunk road. After debating with the driver, Mr Das turned to me to deliver their verdict. It must be that Prafulla Kumar Mahanta was expected. Currently enjoying his second term as Chief Minister of Assam, Mahanta hailed from a nearby village. We stopped for a picnic lunch prepared by the Alpine and just as we were tapping our hard-boiled eggs on the roof of the car, as we had no cutlery, a trumpeting cavalcade swept by consisting of three open jeeps stiff with military personnel, sixteen white Ambassadors, three of which had blacked-out windows and the same number plate, and an ambulance of the Western Front variety. We had just passed a memorial outside a school, to pupils who had perished in the 1978–84 student movement to force Bangladeshis out of Assam. Official figures said the police killed eight hundred young people, but Mr Das said the real figure was higher. Mahanta had led the movement. On the Meghalaya side of the state border, wine shacks crowded the edge of the road. 'Used to be prohibition in Assam,' said Mr Das. 'Thirsty trade other side.'

More than a thousand great one-horned rhinos lived in Kaziranga National Park at that time, two-thirds of the global population (a census taken much later in 2018 revealed that the crash had swollen to 2,413). At the entrance, a pair were drinking at a *beel*, one of Kaziranga's many flood-formed riverine lakes. My notebook records our billet as the 'crappy govt Aranya Lodge, dingy, lots of men hanging round doing nothing except looking at me'. In the early morning the nightwatchman's percussive hawking blended into the fugue of whoops from the moist broadleaf forest. To the north, pink light crept over Himalayan peaks running from Bhutan to Arunachal Pradesh. In the afternoon the air grew close and thunder rippled through the Karbi Anglong hills. The monsoon was coming like a tightening screw, and heat drew sambar deer to the *beels*. A Guwahati television crew at the lodge was making a documentary about Kaziranga, and after interviewing me, the lone foreigner in a park short on bipeds (though full of rhinos), the producer persuaded me to get on an elephant. The Assamese howdahs have three seats in a row, each with a wooden bar in front. When I dropped my lens cap, my forty-year-old tusker, named after a Bollywood star, picked it up and gave it back to me. After that we drove around the park for the rest of the morning, a double-barrelled rifle belonging to the crew security detail in the passenger seat pointing over his shoulder at my head. I had escaped Mr Das for this interlude but as soon as I returned to the lodge he appeared brandishing a copy of that day's *Sentinel* with the headline, HOW BANGLADESHIS GET FAKE BIRTH CERTIFICATES.

I continued clocking up trains. Before boarding, a member of staff ticked off my name on a list. On the seventeen-hour Rajdhani Express a vase of roses and gladioli scented my compartment from Calcutta to Delhi. After three and a half hours on the Ludhiana Express from Delhi to Saranpur (the only train on which the moquette upholstery was dirty) I shared a taxi through what my *Telegraph* friends called 'the cow belt' or 'the Hindu heartland' to Dehradun, where I delivered the speech day address at Welham Girls' School. The head saw me off the next morning on the Mussoorie Express, a train which achieves the feat of covering only 240 kilometres in ten hours. I didn't mind, as Welham had spoiled me with a berth in a first-class ladies coupe. All my rail tickets, still taped into my notebooks, wished me HAPPY JOURNEY in the top right-hand corner.

On one occasion in the mid-1990s I travelled south from Calcutta to the coastal state then called Orissa – now it is Odisha – which is the country's eighth largest, lying south of West Bengal and north of Andhra Pradesh. (The 2021 population of 40 million equals that of Iraq.) The Howrah Mail had conveyed me across the Mahanadi Delta to the capital, Bhubaneswar, in seven and a half hours, and from there it was a short hop to the

coast on the Puri Special. The women working on the new trunk
road out of Puri, sheathed in rose saris, looked fresh even as they
ferried buckets of seething tarmacadam on their backs.

A week in, I was exploring the Niyamgiri Hills in the southwest of the state close to the Niyam Dongar, the small mountain
where the Dongria Kondh people who have lived in the forest for

thousands of years worship Niyam Raja, the centre of their cosmo-
logical world. (I had taken a ten-and-a-half-hour journey on a Puri
to Muniguda train that covered the two long sides of an isosceles
triangle, all of it under a hard blue light that would not turn off.)
Images of India purveyed in the West present teeming scenes,
but few citizens called this part of the Eastern Ghats home. There
was nowhere to stay either, bar spartan government-worker rest
houses ferociously complicated for anyone else to book (though
I did, and stayed in several in Odisha). In the Niyamgiri Hills I
was admiring orchids growing on the banks of a tributary of the
Vamsadhara and had sat down in the scrub to rest. On the nearest hillside a trickle of people made their way down like a dark,
broken stream, all carrying baskets on their backs. Two men, separately, walked up the hill. The pair were Domb, an 'untouchable'
ethnic group. One of them approached a Dongria woman bringing
a single pumpkin down to market. Both stopped. He was holding
a pair of weighing scales. The man and the woman appeared to be
arguing. The woman placed the pumpkin on the scales.

I had read this scene that morning. My heart began to beat
faster when I saw it acted out in front of me. 'Will never forget,' I
wrote in my notebook, 'the reading and the seeing.' In his book
A Goddess in the Stones (1991) Norman Lewis, who had probably
influenced me more than any writer, recounts a trip to the same
part of inland Orissa in which he watched three Domb at Kundili
market set up a wooden tripod with scales. A female Dongria
approached with a load of rice on her back; Lewis had seen her
avoid another three Domb higher up, but now she was caught,
and 'after a brief and hopeless struggle and mewing protest
she gave in, her rice was weighed and she was paid . . . half the
market price'. Seven years later the scene had played out before
me. When Lewis wondered why Dongria didn't 'get together
and put up a fight', he learned that 'the Dombs pay gangsters to
look after them . . . The police are in the Dombs' pay too.' Orissa
has India's third largest population of the people Mr Das called

'tribals'. Lewis was chiefly preoccupied not with inter-ethnic extortion, but with problems of centrally imposed detribalisation. His prose on the subject is characterised by restraint and musicality, his approach by tolerance and detached sympathy. I look up to him now as I did then. He chronicled what was happening to Moïs in Vietnam, to Panare on the Orinoco and to Ayoreo fugitives in Bolivia, and his 1969 article 'Genocide in Brazil', about the persecution of Amazonian peoples, really did change things, which is a good deal more than the rest of us can say as we till the lower slopes. As for the Dongria Kondh, they and their hills sit on a pile of bauxite. Since Vedanta Resources, once a FTSE-100 mining company, announced plans to refine sedimentary rock at Lanjigarh to make aluminium, both the Dongria Kondh and their international supporters have campaigned vigorously and taken a case to the Supreme Court (which subsequently mandated India's first environmental referendum) in order to save the land from depredation and preserve a way of life that has gone on since the battles of the Mahabharata. As *The Hindu* reported of Odisha's Dongria Kondh in 2019, 'Ever since the community turned down a proposal from Vedanta Aluminium to mine the Niyamgiri Hills for bauxite in 2013, it has found itself in the crosshairs of the State government.'

Later that day, hiking a river valley, I stopped at an isolated cluster of shacks. I had been inspecting more wild orchids with a guide. 'So often on journeys through the derelict Indian countryside,' V. S. Naipaul wrote in *An Area of Darkness*, one of the best books ever written about that nation (possibly any nation), 'it had seemed that the generative force alone remained potent, separate from its instruments and victims, men.' But note 'seemed'. The generative force was not separate from its victims. At this place a man engaged my guide in conversation, I assumed in Odia. The guide turned to me. 'They ask if you have medicine. They want you to see a sick woman.' We walked around the back of a hut where a youngish woman lay in the open air on a makeshift bed elevated from the ground. She looked close to death, her skin greyish-yellow and papery, her eyes dark pools sunk into her face. She was uncovered, wearing a dark blue sari, her lower arms wasted. A tall tree shaded the scene. I tasted again the *khaja* I had eaten at lunch.

While I was writing this chapter, a friend's brother, anthropologist and activist Felix Padel, who has been working with the Dongria for thirty years, drew my attention to a sinister development connected with mining in Odisha, one which casts light on the *Hindu*'s 'crosshairs'. The Vedanta Foundation has for a dozen years been paying for Dongria children to attend residential school hundreds of kilometres from the Niyamgiri Hills, specifically at the world's largest boarding school, the Kalinga Institute of Social Sciences (KISS) in Bhubaneswar. The regional government promotes this kind of set-up as the model for tribal education. Shipping five year olds off in the name of learning as part of a general assimilationist ploy to make everyone more like everyone else and weaken tribal links – I've seen that old trick, a colonial legacy, played the world over from New Zealand to Arctic Canada via Chile. But it's 2023. The Vedanta episode is not unprecedented. 'In tribal boarding schools throughout India,' I read, 'children are brought into a different world of values and knowledge.' This op-ed, co-authored in November 2020 by Padel and Malvika Gupta in *Sapiens*, a leading anthropology journal, argues that 'these funding links play a role in undermining resistance to mining on Adivasi [tribal] lands'. It continues,

Taking children away from home to 'civilize' them breaks the circle of life that transmits values, skills, and languages between generations, just as historic North American Indigenous boarding schools embodied the ideology of 'Kill the Indian, Save the Man'. The underlying theory ... [is of] education as a means of speeding up 'social evolution' from a 'primitive' tribal stage toward 'civilization'.

Padel acknowledges that in India boarding-school enrolment is rarely based on physical coercion. However, his research suggests that although factors compelling parents towards enrolment include economic ones, neglect and closure of local day schools also play a major part. Padel had also found that KISS has undertaken recruitment drives in tribal districts 'through a state–corporate nexus that involves police, politicians, and mining officials'. In their report Padel and Gupta show the ways in which many residential schools impose a rigid uniformity 'that is insensitive to tribal ethos, imprinting most children with a sense of inferiority'. Hinduisation means, as one ex-student told investigators, 'In schools like KISS . . . the gods only understand Sanskrit.' In published papers and media reports on the topic, the phrase 'the poorest of the poor' recurs like a tolling bell, justifying systemic adjustment to suit the agenda of the ruling elite. Educational researchers have a name for it: 'deficit discourse'.

Padel met Lewis in Orissa and briefed him. 'The thing I remember,' Padel told me, 'was that he didn't listen to a word I said.' I set this down not just to colour my story with a dab of irony to highlight the apparent gap between words and reality, but because it shows the world as it almost always is – in shades of grey, rather than black and white. So perhaps my companion's assessment of the Mother had been wiser, in the long run, than that of Christopher Hitchens.[1]

And so it goes on, and on and on. Lewis spoke up for the Dongria, Padel and his colleague have taken the baton, and the rest of us must keep speaking, even though we know that one day it will be too late. Perhaps it already is.

I can't remember how I came across Urvashi Butalia, but I think

---

1. In 2000 I visited ninety-three-year-old Lewis at his Essex farmhouse. He was writing a book, later published, about a journey he had made in the late 1920s to Santiago de Compostela. He gestured to a volume on his desk and said, 'I'm doing the research now.' It was the 1998 *Rough Guide to Spain*.

it was through a Radio 3 speech-based programme to which I contributed. Haryana-born and of Punjabi heritage, Butalia, ten years my senior, had recently founded Kali, India's first feminist publishing house. At the Delhi book fair I made straight for the Kali stand and picked up a volume Butalia had co-edited. Most of the stories in *In Other Words: New Writing by Indian Women* (1992) are told in the first person and, according to the author blurbs, are autobiographical. A theme emerged concerning being rich in a poor country. In Shama Futehally's 'Portrait of a Childhood', the narrator discovers that her much-loved ayah is a thief. Several stories dealt with grandmothers and other female ancestors. In the best one, Githa Hariharan's 'The Remains of the Feast', little Ratna smuggles forbidden sweetmeats to her sick great-grandmother. The women were writing about the struggle for identity – who isn't – and when I read reviews of *In Other Words*, I learned that many of their issues reflected a discourse surrounding Indian literature in English. Here's one:

> As with all Indian writing in English, there is a self-consciousness that marks these stories, a self-consciousness often discernible in terms of the slightly apologetic nature of several of the narrators, who are acutely aware of their upper middle-class status in a country where class distinctions are easy to locate and can often be seen as marked by the use of English as the medium for creative expression.

I wonder if self-consciousness doesn't mark all stories, whoever writes them. One reviewer noted it was 'significant' that Indian women writers felt the need for the confessional 'a full fifty years since Ismat Chughtai'. The prizewinning Chughtai (1915–92) wrote in Urdu about women's inner lives. Governments banned her books, but she wrote for early Bollywood. Her short stories should be reissued. My favourite, 'Chui Mui' ('Touch Me Not', 1952), involves a woman who had not 'grown up with ayahs' but

had married the Shehnai of Bhaijan 'and become tender as a swollen wound', bejewelled so 'a mole could not have peeped through'
her multitudinous amulets. (All quotes from M. Asaduddin and
Sunrita Basu's translation.) When the protagonist falls pregnant –
Chughtai often writes of the pitfall of equating motherhood with
womanhood – her relatives 'stitch diapers' in a frenzy of proleptic
excitement. The reader glimpses, early on, that the gilded cage
is tarnished. The woman lives in fear that another miscarriage
(she has already endured two) will be her husband's 'ticket to
second marriage'. Begum in-laws decree she must deliver her
baby in Aligarh, two hours from their Delhi home. The family sets
out in a private carriage on the Toofan Mail, and here Chughtai
produces one of the best passages ever set on an Indian train. A
poor woman pushes her way into the carriage, pregnant herself,
and the powerless family watch her give birth on the floor and rip
a strip off her dupatta to tie the umbilical cord. The 'flower-like
Bhabijan [the pregnant wife of the Shehnai on her way to Aligarh]
felt so unnerved' that her unborn child 'got cold feet and wilted
away before its entry into the world'. What a film it would make.

People have called Chughtai India's Simone de Beauvoir. I see
de Beauvoir as France's Ismat Chughtai. Or is she France's Pandita
Ramabai (1858–1922), who also should be better known? Two
generations before Chughtai's electrifying scene in the railway
carriage Ramabai had campaigned for women's rights in Marathi-
speaking western India (now Maharashtra) and nationally,
mustering female delegates and standing herself for Congress in
1889 as well as setting up homes for vulnerable women and promoting female education. Hers was not a lone voice: Maharashtra
had its women's press a century ahead of Virago. Feminists
showed a radical approach towards empowerment by creating
their own narrative. In 1905, twenty-five-year-old Bengali Rokeya
Sakhawat Hossain (also known as Begum Rokeya) wrote, in
English, the feminist fantasy *Sultana's Dream*, set in Ladyland.
In this paradise men live in a harem and women control the

weather by trapping solar power. When Begum Rokeya's narrator hears that men's brains are heavier than women's (and therefore, it follows, more effective), she points out that elephants' brains are also heavier. Ladyland's women decide that the working day can be cut from eight hours to two, as when men were in charge they spent six hours smoking. The pioneering Rokeya went on to found Calcutta's first school for Muslim girls.

Promising beginnings, on which I would like to write more. Here I will say only that over my six decades I have seen this whole empowerment feminist idea defanged in the West. The call is to 'lean in'. Women have witnessed the depoliticisation and even corporatisation of feminism; if we all lean in, who is leaning out to build support for others? By others I mean the vulnerable, the economically disenfranchised, the young. Has a political frontier not been dismantled? Looking back to my twenties, I recognise the subsequent erosion of collective feminist will. The god of choice has risen in its place. Trickle-down feminism has not worked, just as trickle-down economics failed. Indian feminists identified this trend almost forty years ago in 1984 when feminist scholar Gita Sen, subsequently consultant to the World Bank and the UN and now at Harvard, formed, with other women from the Global South, a then Bangalore-based collective centred on political mobilisation. DAWN (Development Alternatives with Women for a New Era) deployed the term 'empowered' to introduce bottom-up feminism, arguing that grassroots organisations 'could be the catalysts of women's visions and perspectives' and contribute to a mandate for structural change. You can't help wondering what has gone wrong in Ladyland.

Chiaroscuro had emerged in my personal Ladyland: six months on the road and six months in Mornington Crescent. Garlic no longer carried a hazard warning in Britain, but I remember a big Sainsbury's opening in Camden Town in 1988 on the day I moved into my flat. When I went in to stock up on basics, known then as litre cartons of red wine, a woman was scrutinising the ingredients of a tube of tomato paste from a shelf above one of the freezer aisles. 'Look,' she said in exasperation to her husband, 'there's garlic in *everything*.' Sometimes I rented out my flat when I went to India. Wherever I was, I wrote a lot of features. One involved belly dancing, and Willie Rushton drew me at it.

Sometimes I took on editing assignments to fatten the slender income from my own writing. I enjoyed editing Norma Major's two books. For her history of Chequers, she and I sequestered ourselves in the Buckinghamshire house itself. A policeman flung the door open when I left on my daily run. I swam in the unheated indoor pool. Mobile phones did not exist, so when I rang anyone I had to go through the Downing Street switchboard. At about the same time I had a job writing obituaries for an annual directory (it did not record that year's deaths, but was working its way through the previous decade and filling holes in the decades before; these late entries were then added as appendices to the next printed volume). I had to produce three a day, and the only one I remember is Jennie Lee. Thatcher had left office during this Indian period. I was in London the day it happened, staying at a friend's place until a tenant's lease came to an end in Mornington Crescent. Like everyone of my vintage I remember the watery-eyed shot in the back of the taxi. John Major was a benign presence in Number 10 compared with what we had lived through, and I knew him personally to be a decent, likeable man. But a Jennie Lee was not waiting in the wings despite the hope I mentioned in the previous chapter. I was only thirty, but coming of age under Thatcher had instilled caution that a new dawn might be glimmering just beneath the horizon.

I had been battling on with feminist theory. Judith Butler produced *Gender Trouble* during the Indian phase. In that book Butler suggested that, consciously or not, people are citing conventions of gender when they claim to be expressing their own interior reality or even when they say they are creating themselves anew. I did not say I was doing any such thing as 'creating myself anew', but was I 'citing a convention of gender' as I took off to observe the world? Young men I met on the road were more confident – a generalisation I make without qualms – but I'm not sure they behaved altogether differently. It was not, however, insignificant that I was able slip into women's spaces like the kitchen or the hammam or the marketplace. I wanted to make my status as woman part of the stories I told. So often, as I tried to work out what I was doing in my work in this regard, I longed for female role models all over again. Publishing had progressed since Penguin's 1962 *The New Poetry*, a work, as I mentioned in the Introduction, which showcased only men's poems. I did what I considered my duty, inventing and co-editing the 1997 *Amazonian: the Penguin Book of Women's New Travel Writing*. I hoped it might explore the women's room in the notoriously raffish open house of travel writing.

I was in Bangladesh in 1996, my first long post-polar trip. (My time on the ice had more or less put an end to the Calcutta period.) Antarctica is a hard place to evict from your head. I felt in my heart and still do that I am a generalist, and although I could have gone on writing about little wooden ships in the pincers of a floe for ever, and paid the bills through the egregious hardships of those who had gone before me, it was not what I wanted. My plan was instead to spread the travel thinner and, specifically, not to become a self-proclaimed expert. To make sure the Antarctic break was a clean one, I selected, as its polar opposite, the most densely populated nation on the planet. That was Bangladesh.

One day, in the southwestern city of Chittagong, I took a painted cycle rickshaw out of town, north to the university. Someone in Dhaka had recommended a young man who studied medicine there (this was before the foundation of Chittagong's separate medical university). 'He speaks good English,' said my informant, 'despite never having left *Banladesa*, enjoys meeting foreigners though none ever come to Chittagong, and has an interesting take on his country.' I had the name of the campus written in Bengali literally on a Hollywood cigarette packet, the driver knew it or said he did, and off we went, an unlit kerosene lantern swinging from the undercarriage and occasionally clonking an obstacle in the road.

Chittagong, second city and proud first port, sits on a natural harbour on the Bay of Bengal west of the hill tracts that separate Bangladesh from Myanmar. It was January, the air balmy in the way spring there lasts half a year. Huts teetered and spilled onto the road as my driver pedalled into the sun, and everywhere boiled over with people. Men with angular legs squatted mirror-lessly lathering their faces. In one street a string of women sat in knots unravelling acrylic sweaters. Next to each knot, a small girl piled balls of yarn into colour-coded ziggurats. The mosques at that hour were silent. Chittagong, like the whole country, is overwhelmingly Sunni, and Bangladesh the third-largest Islamic nation in the world.

Students were milling outside the university, some resting on the handlebars of enormous bicycles in varying states of decrepitude. Most were talking, holding folders and books and files in the way students used to everywhere until the microchip landed. Most of the men were wearing kurtas and trousers in a neutral palette. A third of the young women wore salwar kameez, a third saris and a third western dress. Nobody paid attention to me, but once I had entered the lobby and a student asked if he could help, people began crowding round. A western visitor constituted a significant event in an ordinary university day. I produced another portion of the cigarette packet marked with the medical student's name.

Bangla is the only language I know which can be soft and loud at the same time. It rises and falls in such a tender flow that the speaker can raise their voice without introducing a harsh timbre. Much loud-soft talk ensued in the light-filled university lobby. Students scattered up the wide staircase to the right. The rest of us exchanged smiles. A group of young women gathered on the left around a tall girl giving out pieces of paper. Sunlight fell in steeply angled rays.

Shortly a thin man of about twenty in a biscuit-coloured kurta stood before me. 'I,' he said, 'am Devesh Khatun.'

I am more surprised now than I was then that this implausible search had worked. Devesh was indeed keen to talk. There was no student canteen, so at first we sat on a wall in the campus grounds, joined by a fluid assortment of other students. Conversation ranged from events in London to a recent train derailment. But it is the evening I remember most clearly. Devesh had invited me to his student flat. I travelled in a cycle rickshaw, Devesh instructing the driver in detail before jumping on his own bike. It was late afternoon by now, and on the banks of the Karnaphuli stray dogs scavenged with lassitude. Rickshaw cyclists were lighting their lanterns and the whole of Chittagong smelled of kerosene. When we jolted to a stop outside a quietly decaying three-storey brick house, of which the ground-floor windows stared out glassless like eyes gone cold, Devesh was waiting. 'Welcome,' he said, '*Shagata*.'

I had been learning Bengali at night classes at home, but in situ I understood nothing, and as each week in the country passed, I was able to stammer fewer of the key words I had learned. Reverse osmosis was taking place.

We walked up pockmarked stairs sporting neither handrail nor banister and on the second floor entered, through a plastic curtain rather than a door, a very small flat with a very small roof terrace. Off this concrete terrace an extremely small woman squeezed into a tiny scullery was bending over and stirring a large pan. An unidentifiable aroma infused the brick terrace. Chittagong, spread out below, steamed off its day.

Devesh introduced me to three flatmates, also medical students. We sat on thin cushions on the terrace floor. The four young men lived on a slim monthly *taka* allowance from their parents yet were able to employ the diminutive cook, who came in every day. She remained silent, and slipped away after she had washed up the dishes from our supper in the exiguous scullery. The five of us sat talking as the light failed, about the *mahajons*, the Bengal-style mafia which runs half the country, and the stock exchange due to open that year in Chittagong, consolidating the city's role as a financial hub on the Bay of Bengal. The other three students spoke reasonable English, but we relied on Devesh to supply nuance. He had learned through the BBC World Service and one textbook which he had read thirty-three times. His uncle had owned a short-wave radio and when he died, his widow had given it to Devesh, the most bookish of his generation of cousins, and he had mummified it with Elastoplast as bits fell off. We discussed the fact that in Bangladesh both the prime minister and the leader of the opposition were women, but that both had taken on almost hereditary roles, the PM as daughter of Sheikh Mujibur Rahman 'founder of the nation' (he was assassinated in 1975), and the leader of the opposition as the widow of Ziaur Rahman (President Ziaur, killed in 1981 not far from where we were sitting cross-legged). For the students, let alone ordinary women imprisoned by problems of their own, these figureheads were remote. But the young medics discussed them gravely.

During that kerosene-scented Chittagong night Devesh said, when one of the others spoke of the medical knowledge they had already acquired (they were all fourth-year students): 'Oh yes, we know the answer to a diagnostic neurology question. We write about MRIs. But we have never seen an MRI machine. And we will never see one, if we practise in Bangladesh till we are fifty, which will be in 2020.'

In 2018 the government changed the city's name to Chattogram to reflect its Bengali pronunciation. At the time of writing the city's stock exchange has a market capitalisation of US$38 billion. I do not know if the students ever saw an MRI machine.

I travelled down the Gangetic delta on a Rocket paddle steamer, a twenty-four-hour trip from Dhaka's Sadarghat to Mongla Khulna. The Ganges officially becomes the Padma when it enters Bangladesh and the Brahmaputra is renamed the Jamuna;

halfway down they fuse and become the Meghna. This river moves 400 million tons of earth a year and regularly rearranges the map. As for the magnificent seven remaining Rockets, built in Calcutta in the 1920s, their paddle wheels were newly powered by diesel rather than steam, but they embraced the modern age with dignity. It was on the top deck of the Rocket, observing standing oarsmen sunk within tightly wrapped shawls in flotillas of small boats emerging from morning mist, that a turbaned waiter bearing tea and Bengal orange cake first addressed me as 'Sir'. I liked this very much. It was not that he thought me a man. It was his wish to use a respectful English greeting.

Mostly, though, I took Bangladeshi trains (the tracks ended then in Chittagong). I spent many days looking out of carriage windows, but wonder now why I noted so little about the built environment. I can't remember, in Shillong or Mysore or Cochin or any of the places I loved deeply in India, thinking about the architecture all round me. Nor can I find even the briefest comment on the topic in my notebooks, bar a few references to leprous stairwells. What was the matter with me? Did I not see? It was several decades later, on my first trip to Sri Lanka, that the subject pierced my elephant hide.

I had hiked to the summer palace of King Kassapa on top of a fearsome rock that rises vertically from the central plains. The knuckled masonry of the Sigiriya royal complex has fought bravely, for more than a millennium, against the incursions of the jungle, though a tropical languor overlaid the scene, sweet with desuetude. The summit site, built in the fifth century CE, covers 1.6 hectares and includes drinking-water tanks, a swimming pool for five hundred concubines (presumably they didn't all get in at the same time), a throne room and a parliament hall. All of it, spread out on that high rock, overlooks fertile plains, misty in the early morning and humming with green bee-eaters. The illegitimate Kassapa, who had murdered his father, wanted an impregnable fortress with thousands of soldiers garrisoned at its foot. When the monsoon hurried in, the court moved to another palace in the water gardens below.

Fifth-century architects considered the relationship between a building and its environment. Much of the eastern moated precinct of Sigiriya mirrors the wide limestone belt on one side, and the symmetry pleases the eye. At the outskirts of the site, narrow entrances give out, in sudden shock, onto wide-open spaces with water gardens that echo the lines of the rearing rock beyond, and onto another moat overlaid with mats of blue lotus. Halfway up to the palace, brick-built lion paws either side of a lost head revealed that visitors once entered through a leonine mouth. A man had attached himself to me as 'guide'. I didn't want a guide, but was in too pleasant a mood to be rude. 'What kind of bird is this?' I asked as a small warbler danced around the stonework. 'A brown one,' he said.

Shortly after my ascent of Sigiriya, I travelled down to Galle on the southwest coast, through paddy fields and banana plantations striped with filtered light. Pyramids of hairless coconuts rose at the roadside stalls. The heart of Galle lies within a seventeenth-century Dutch fort (built from coral and granite on the site of an earlier Portuguese one) on a promontory spearing the Indian Ocean. In the company of Channa Daswatte, a Sri Lankan architect with a practice in Colombo, I toured the modernist buildings inside and outside the fort. The road signs were in Sinhala, a language with a script that could challenge Assamese; in the straighter Tamil alphabet; and in English. In the rectangular grid of streets, outside a shop marked SINGER, a man treadled at a sewing machine. Daswatte hurried on, eager to introduce his idol Geoffrey Bawa (1919–2003), high priest of tropical modernism. Bawa designed the dazzling Jetwing Lighthouse we were looking at north of the fort. He conceived the purpose-built hotel as a link between traditional architecture and modern Asian minimalism. The sixteenth-century Portuguese landing at Galle plays out in brass and bronze statuary on the main staircase, a spiral heading up to a well of light and a coin of peacock-blue sky. The whole design demonstrated Bawa's integration of architecture and landscape: the polished concrete floor of the main lobby appears to flow directly into the ocean, while functional features – swimming pools, for example – reflect the natural shapes of their surroundings, from banyan trees to clifftops.

'Unlike India and Bangladesh,' said Daswatte, running his hands through a stook of streaky grey hair, 'Sri Lankan architects did not, at the advent of modernism and independence, make a clean break with the past.' I thought back, as if for the first time, to the buildings of Le Corbusier after he was drafted into Delhi, or to the Dhaka Parliament of American Louis Kahn. 'My predecessors,' said Daswatte, 'grafted modernism onto what was already there, and adapted it. They were concerned with a building's relationship to its landscape, like King Kassapa.'

'To feel useless or helpless is the way most people feel,' Martha Gellhorn wrote after a stint viewing corpses in Vietnam,

> ... and it is a bad way to feel but also an excuse. If you can do nothing to change events or to rescue your fellow men, you are free to lead your own life, and living one's own life is always more pleasant than the exhausted scrabbling role of a responsible citizen.

Of the patients in the Qui Nhon hospital in the old provincial capital in the south Gellhorn says, 'We big overfed white people will never know what they feel.' True, and where does that lead the itinerant scribe? Gellhorn had a pass as she was a war reporter, a real and important job in which one was fighting all the time on the right side. I am confident of my own insignificance, both on the road and on the page, but something unresolved lingers in the inability to act. I did nothing about the inequality I observed, in Galle, in Calcutta, anywhere. 'Leading one's own life' is indeed more pleasant. I put life into boxes to facilitate the arrangement. When I look back over the years described in this book, I perceive my life falling neatly into a 'Here' box and a 'There' box. 'There' was home, a place of stability where everything was the same when I staggered back. Of course, it wasn't, but that is the trick memory plays. In fact, every time I came back something had changed. The pound transitioned from a note to a coin, television programmes kicked off at breakfast time, Tango introduced its blackcurrant flavour. And it never ends. Even the pound coin isn't safe: in 2017, shapeshifting like an amoeba, it developed twelve sides while I was in Sichuan. Four years later I returned from Corsica to find 'dark berry' Tango on the shelves. What next? 'Here', on the other hand, was wherever I was, where I was truly alive, where I did my thinking, insofar as I did any. This latter can't be true either; all the ingredients of thought were macerating while I was at home paying the gas bill, mending a leaky roof and buying crampons. Someone once wrote of cognitive anchors – the things you see through the periscope to reassure yourself that the world is still there. Looking back, I can't work out where the periscope was: in Mornington Crescent, or out beyond the blue. I am talking about the chiaroscuro I mentioned earlier, a state of being I have noticed in many travellers. Freya Stark, among my top heroines, sought light and shade in abundance. Her friend and mine, Colin Thubron, wrote that she was 'Nomad and social lioness' tacking on, 'emotional victim and myth-maker. It was rare to leave her company without feeling that the world was somehow larger and more promising.' Unlike Colin, I did not know Stark, who lived from 1893 to 1993, but I did go to her memorial service in St James's Piccadilly. I can't remember why – I think I had a commission to write something, or it might have been one of the obituaries. It was a thrill to get close to her, even in death. 'Surely,' she wrote, 'of all the wonders of the world, the horizon is the greatest.' She covered mostly, on foot and on the page, what we used to call the Middle and Near

East – Turkey, Iran, Iraq, Syria, Egypt, Afghanistan. On her third
visit to Turkey she took the same route as Alexander the Great
twenty-two centuries earlier, 'through lonely mountain passes,
past remote ruins and along the pirate shore where country
people respond gladly to informed curiosity.' A female academic
began a lecture on Stark like this: 'It is said that a man who travels
will document his observations whilst a woman writes about her
external and internal journeys: her experiences, her delights, her
hopes, her fears, and of those she meets on the way.' I am not sure
if that is true any more, or if it ever were true. I think though
that Stark had an acute sensitivity to the role of memory in a
woman's emotional life. 'In smaller, more familiar things,' Stark
wrote, 'memory weaves her strongest enchantments, holding us
at her mercy with some trifle, some echo, a tone of voice, a scent
of tar and seaweed on the quay.' That quayside tang is indeed
more agreeable than the 'exhausted scrabbling' of a responsible
citizen. Stark's consuming fascination with the region she loved
began on her ninth birthday when her parents gave her a copy
of *One Thousand and One Nights*. 'The essence of travel is diffuse,'
she wrote. 'It is never there on the spot as it were, but always
beyond: its symbol is the horizon, and its interest always lies
over that edge, in the unseen.' Lawrence Durrell called Stark
the 'Poet of Travel' and for once he was right. A piece in the
Chandigarh-headquartered *Tribune* recently compared Stark's
horizon comment with that of the Urdu poet Ahmad Nadeem
Qasmi. *Meri manzil ko ufaq paar bataane waaley/maine dekha hai,
ufaq taa-ba ufaq koi nahin.* 'Hear from me, you who told me that
my destination is that horizon: I have been there, and seen for
myself, one horizon after another. There is no final one; the end
is simply not there.'

4

The Storming Army

Africa

These momma faces, lemon-yellow, plum-purple,
honey-brown, have grimaced and twisted
down a pyramid of years.

MAYA ANGELOU, 'Our Grandmothers'

I didn't write the Choo Choo book because I had a baby. Throughout my late thirties and forties, I travelled with my children, notably in Kenya, as for a period I was hunting Denys Finch Hatton (1887–1931). He was a hunter himself, and I had signed up to write a biography. On the road I was no longer the single, nubile female who made people either uncomfortable or predatory. I was suddenly fabulously benign, an ambulatory signal that all was right with the world. And of course, as the children grew I began to see places through their eyes. As the Urdu poet wrote, there was no final horizon.

The actual world order shifted on its axis too. In between the birth of my boys, 9/11 happened. Everything connected with travel changed. A vein of tension began to flow which my generation of boomers had not known. The scent of seaweed on the

quay carried a note of aviation fuel. Fortunately I did not depend
on the travel pages for income. By this stage I knew that the long
form suited me best – I was a marathon runner, not a sprinter. I
enjoyed the sustained immersion involved in book-writing. With
three volumes behind me, I had begun receiving regular letters
from readers too. It was a thrill to sense engagement, after years
when only I knew the dynamics of the page. Naturally, my correspondents were sensitive, intelligent readers with an eye for a
good book. Mostly. I cherish a missive typed on small sheets of
blue bond paper which begins, 'I have waited years for a good
book about the Antarctic. Yours is not it.'

I knew what I wanted to write first instead of *Hindu Choo Choo*,
but I had to work at getting permission. Making my way through
the polar shelves during the Antarctic period, I had discovered
Apsley Cherry-Garrard's *The Worst Journey in the World*, which
appeared in the magical 1922. The author had marched part
way to the Pole with Scott, and the following spring was among
the party which crawled into the last tent to find three frozen
bodies. Cherry, as he was universally known, had tried to redeem
his losses on the page. 'If you march your Winter Journeys,' he
ended his book, 'you will have your reward, so long as all you
want is a penguin's egg.' Alone of Scott's protagonists he did
not have a biography, and I thought he deserved a place on the
shelves, as his volume will outlive the next ice age. Angela, his
widow – they married in 1939 when she was twenty-three – had
always withheld permission (which meant permission to quote,
including from the archive at the Scott Polar Research Institute in
Cambridge). She said Cherry would not have wanted a biography.
I put my faith in the importance of the historical record rather
than the wishes of the deceased and took the train regularly to
Sudbury in Suffolk, leaving my first baby in the care of a nanny,
to build up a relationship with Angela. After two years she rang
to say, 'Will you write Cherry's biography?'

To write a life is to take a journey – a serpentine one through dark woods and across blue seas to a bad ending. Cherry was my first Life. Michael Holroyd, master of the biographer's craft and the writer I would like to be, told me that only two criteria matter in the selection of subject: early death and good handwriting. Cherry, born under Lord Salisbury, lived to see Sputnik – 'We shall reach the moon in a thousand years,' he had written – and his handwriting wasn't bad, but I learned the emotional truth of Michael's words when I reached chapters I longed to call 'And So the Years Passed'. I enjoy biography-writing. It is transgressive – inhabiting someone's life, or trying to. I feel at home in the biographer's space where the public meets the private. I enjoy excavating sources, searching for scraps of illuminating detail among huge heaps of negligible things. I like the fact that the elusive, unreliable and flame-like nature of primary material is not unlike the human spirit: capricious, contradictory and inconclusive. I love setting figures in their historical and social context. And I like the way they don't answer back, because they're dead. Every fortnight, while I was writing *Cherry*, I took the train to Sudbury again and walked past Gainsborough's house to Angela's own. She had watery blue eyes like Sophia Charlotte and made me *coeurs à la crème*, my favourite. Besides a masterclass in the fallibility of memory, the experience lent history fresh texture: I was talking to someone whose father-in-law was born in the reign of William IV.

But in the period after *Choo Choo* had pulled out of the station and before a green signal flashed for *Cherry*, I had to do something. I had nervously told my late editor Tony Colwell over lunch at an Italian restaurant on Vauxhall Bridge Road that I was going to be delivering a baby, not a typescript. After his effusive congratulations I said, 'But there's a contract for a book.' 'Bugger the contract!' he yelled as a startled waiter appeared to take our order. It would not have occurred to me then or at any time not to work. I still had the old short-length writing commissions, and continued with book reviewing, which I had begun at the start of my career and have never abandoned. You learn a lot from reviewing, and it complements the travel writer's craft; as Jonathan Raban wrote, 'It's as hard to bring a book convincingly to life on the page as it is a landscape.' (Raban does though remind us that the literary editor in Cyril Connolly's *Enemies of Promise* is called Mr Vampire.) I could take a baby on some assignments. Before Wilf, my first little boy, was two months he had been to Dublin and Cadiz and on a book tour to New Zealand. My flight out to the latter landed in Auckland at six in the morning. Half a dozen authors were on the plane, as the tour began at a book festival. A row of publicists waited to greet us and when the Random House representative spotted the baby carrier in my hand I heard her say, 'There's mine.' But odd pieces weren't enough to keep us in nappies, and in desperation I sold a series to *The Sunday Times* which I called 'Travels With My Baby'; they changed the title to 'Have Baby, Will Travel'. The series was appalling and I hope nobody ever read it. But I was able to travel with Senior – who was of course not yet Senior, he was Only – and pay my half of the bills which regrettably don't stop coming in when you have a baby.

One of the first stories in this lamentable series described a trip from Sydney to Manila on the *QE2*. I had already had one experience of that ship, as in 1995 I accepted a commission from a newspaper to travel transatlantic with my father. I did this because I knew Dad would love it (he did), and because I wanted to write a piece called 'The Old Man and the Sea' (I did). Dad and I flew to New York and stayed two nights at the Four Seasons, where he lay in bed pressing the remote control which opened and closed the curtains. Liberal twits like me go on about the ghastliness of cruising, but I didn't mind it, despite the fact that on the Atlantic passage the traveller benefits neither from weather nor scenery. I used to lounge on deck twelve wrapped in a horse blanket while the steward ferried beef tea. On the later cruise I took the baby ashore in Papua New Guinea where he was a hit. Blond hair was his calling card in those days. On board, a distinguished red-top newsman was at my table in the Princess Grill. Although I never took Wilf to dine but booked a babysitter from among the crew, the grizzled hack took every opportunity to trumpet his view that infants should not be 'allowed' to travel. At the time I was vulnerable on a number of fronts, but not that one: in general I reckoned that old reactionaries would die shortly and their opinions with them. You choose your battles, don't you?

It is easy, if you don't mind not sleeping, to travel with a baby, as he or she wants nothing except to snuggle close. I took my second son on the road with the Sámi of northern Sweden as they brought their reindeer down from summer pasture. Sámi, also called Lapps, live in an arced band across northern Scandinavia and western Russia. They are no longer nomadic or even migratory: they bring their animals up and then take them down as the seasons unfold. The story I was pursuing focused on the Sámi struggle to maintain traditions and values in a modern world poor at accommodating divergence. This by now had become a regular theme of my work: I could give marginalised people a voice. As a working mother dragging a baby round on assignments I felt marginalised myself.

I was feeding the baby myself too. It was minus thirty in Sweden north of the Arctic Circle. I had a kind guide who had children of his own, and on the first day he produced a sheet of foil to stick down my bra. This was to facilitate heat transmission when baby suckled. I never got on very well with the foil, and dreamed of the Princess Grill, where the waiter stood discreetly behind my chair and murmured, 'I do have some beluga, madame.' A picture of Reg on a sledge made the Christmas cover of the *Telegraph* magazine, but all I remember is being tired. Once Wilf, the first baby, began taking solids I had become an expert on portable sterilising equipment, including gear that did not require a power supply. Years later, after two small boys had become robust walkers, our all-terrain collapsible buggy was

too knackered to be passed on to another mother: we had worn it out. But I am too jaded to get excited about kit, even a green babyjogger soaked in happy memories. In extreme environments a certain type of man goes on about gear all the time. They put the kit before the horse. And still women work in the Antarctic with frozen stock cubes in their parka pockets.

In 2003 I took both boys and their father to the Kenyan coast and learned to cut malaria pills into quarters. We rented a house a hundred kilometres north of Mombasa, near Watamu. The elderly white man who owned the property had recently died, and it had not been fixed up as a rental. Termites had eaten the books, mostly hardbacks about settlers who came in the fifties, presumably when our deceased landlord had arrived. The two-storey house over-looked a band of forest which in turn gave onto the Indian Ocean. A houseguy who came in every day was terrified of us, and who can blame him, and he wrung his hands when the roof fell in over one bedroom. Mariam also came in to clean, and when I asked her if she would be able to look after the baby for an extra fee she indicated that she would be delighted. Reg was eleven months old, and she put him in a sling on her back.

These were the Finch Hatton years. He was an English aris-tocrat who had turned his back on all that and set up in British East Africa, as Kenya was, first as a trader, then, in the 1920s, as a safari guide. He was Karen Blixen's lover, and Robert Redford immortalised him in the multi-Oscar-winning *Out of Africa*, based on Blixen's book. (Unlike Redford, DFH was bald as a bil-liard ball.) At any rate having signed up to write his biography I was keen most of all to paint the landscape against which this mysterious wanderer lived his short life (he pranged his plane into the green hills of Voi when he was forty-three, so this time I had taken Michael's advice). Finch Hatton had built a house on the coast not far from Watamu. I was astonished, when I visited, at how little the immediate surroundings had changed. In fact, they had not changed at all. DFH had bought the land at a place called Takaungu in 1927 and built the house the following year on the ruins of an Arab settlement. An Indian businessman in Mombasa owned it now though seldom went, and I had obtained permission to visit. Situated fifty kilometres north of Mombasa between two deep tidal creeks, it remained an isolated spot, reached by a rutted track off the Malindi road. The boys ran around the house, a low Moorish dwelling built from plastered coral blocks. At the front an arched colonnade faced the sea and a low wall ran between the arches, broken by steps that curved to a sandy beach. A pair of carved doors with worked brass-copper bolts opened from the tiled colonnade into the dark, cool interior, and DFH had designed that part so that when the moon was full it shone through the house. At those times Tania, as he knew Blixen, wrote 'the beauty of the radiant, still nights was so perfect that the heart bent under it'. When the tide was out they could walk over the beach picking up shells before retreating to a cave until the water returned and the caves filled up, 'and in the porous coral-rock the sea sang and sighed in the strangest way'. The long waves came running up Takaungu creek then like a storming army. After a three-month safari Finch Hatton would walk down the steps and bathe, as we did, and like him my children went to the stone walls at the south end of the beach and caught sand crabs and brittlestars in the coral pools. In the other direction, in a band of gathering green, we watched the descendants of the *usambara*, a coastal tropical butterfly, that Finch Hatton had seen shiver in the orange flowers of *Cordia africana*.

Senior celebrated his fifth birthday in the Watamu house. We went to an animal sanctuary – his choice – and he sat astride a turtle that was three hundred years old (or was it three thousand?). The next day, the dog attached to the house, I assume still mourning its owner, bit Junior's face. I stayed calm – you do, I think, in the heat of panic – while we went to a clinic; established from vet records that the hound had received its vaccines; and talked to a paediatric friend in the UK. All was well, and the boy, now six foot five, doesn't have a scar, but that night, had I had access to the internet, I would have booked up for next year and every succeeding year at Pontins on the Isle of Wight. What kind of idiot was I, exactly?

Three years after that first visit to Kenya I took Senior back on safari for a newspaper piece (I was *that* kind of idiot). We stayed in a tented camp. It was my good fortune to have given birth to a twitcher. I don't know where it came from. Wilf read only books about birds and quickly became an expert spotter, so I tried to pick places with an interesting ornithological population (this turned out to be everywhere). Of course, I began to learn about birds myself. Wilf was passionate about animals too. He cried himself to sleep in his cot bed one night in the Serengeti because tourists in the other jeep had seen an anteater, and we had not. I was a hopeless parent in every way and took a couple of serious wrong turns, but I was able to do that one thing – show him a slice of the natural world. Once we went across the Arctic Ocean in a Russian icebreaker and when he saw his first polar bear he said, 'This is the best day of my life.' (He learned a lot of unsuitable Russian phrases on this voyage as the crew adopted him.) Even in the south of France, where I did a house swap, he spent hours examining grass snakes.

Light and shade both took on a fresh tint; it was never quite the same chiaroscuro as it had been, and whenever I pass the end of Mornington Crescent on the top deck of a 24 bus, I glimpse those fugitive days when I could shut the front door on my English life.

With children in tow I had even fewer role models. Ten years ago I wrote this paragraph in a piece for an anthology, and I can't say my attitude has changed:

> The history of travel writing reveals few mothers. Have a baby, and you lose your passport to the magical world of anonymity, impulse and sleazy bars. The famous fathers of the genre usually had that most valuable travel accessory, a wife who stayed at home minding the brood. (Forget the multi-outlet electrical adaptor. Get a wife!) *The Great Railway Bazaar: And the Kids Came Too.* I don't think so.

All the pieces about travelling *en famille* in magazines and travel pages, some of which I wrote myself, were, it seemed to me, about showing off. Either that or they indicated devalued currency, like anything tagged both 'children' and 'travel'. I feel the same about any phrase with 'family' in it. A restaurant with 'family' as its selling point means it's crap. It's a trade-off – we'll have your children in here and you eat our filthy food. As a writer, I was afraid the trade-off would be covering issues that didn't interest me – that weren't *important*. I still thought I was writing about important things.

Dervla Murphy pulled it off. She wrote travel books while on the road with her daughter, then the daughter grew and left home and left the books too. Murphy resumed *sola*. I love that cycle (and she often went on a cycle). She was born in 1931 in County Waterford, and for me she was always there. I met her several times: once we did a joint event at the Royal Society of Literature in London's Somerset House. She was brilliant, and all I had to do was sit on stage and tell a few stupid stories so the evening wasn't embarrassingly unbalanced. Murphy was the opposite of the successful men in her field in that she disliked public events – half the men say they don't enjoy standing on their hind legs, but they always seem to relish it really. On this occasion Murphy was in town for a week to promote her latest volume. Halfway through the week she had a day off from speaking engagements and flew home. Her publisher managed to get her back; she said she had gone to feed the animals.

For her tenth birthday Murphy had received a second-hand bicycle and an atlas (a pattern is forming here. Give kids significant travel gifts at nine or ten! Or don't, if you want them to live a safe life!) In 1965 she published *Full Tilt: Ireland to India with a Bicycle*. She had pedalled through Europe in 1962–3, the worst winter for decades. My brother was born in Bristol that December. Central heating had not appeared in our habitat and he did not leave the house for six weeks. Or so I am told. I suppose someone took me out. Murphy though cycled on and liked Afghanistan so much she said she was an *Afghanatic* and

,that the Afghan was 'a man after my own heart'. The bike, an Armstrong Cadet she called Roz for Rocinante, Don Quixote's horse, was a character in *Full Tilt*. Cycling was Murphy's selling point as a travel writer – a useful one to have, as bike books do well. Murphy broke the habit in her fourth book, *In Ethiopia with a Mule*, when she walked from Asmara to Addis Ababa with the beast alongside her carrying the gear. But it was the years with her daughter that most interested me. She took Rachel to India and South America, as well as Africa. Their last trip was on horseback through Cameroon, where Dervla was frequently mistaken for Rachel's husband (*Cameroon with Egbert*). They come across as a team, which is what Murphy says, at the outset, they had become, over the years. On travelling with a child, she wrote:

> A child's presence emphasises your trust in the community's goodwill. And because children pay little attention to racial or cultural differences, junior companions rapidly demolish barriers of shyness or apprehension often raised when foreigners unexpectedly approach a remote village.

Alone again, she set off for South Africa (*South from the Limpopo*), Kenya to Zimbabwe (*The Ukimwi Road*), Rwanda. Then, three granddaughters walk onto the pages of a book about Cuba. Murphy turned seventy-four on that trip; they were ten, eight and six, and the youngest told everyone, after Murphy furtively opened a can of Bucanero early one morning, 'Nyanya's having beer for breakfast!'

Throughout her career Murphy has tackled things that matter: Northern Ireland, nuclear proliferation, tribal displacement, the impact of AIDS in Africa, post-apartheid South Africa, genocide (in *Visiting Rwanda*, one of her strongest). In *The Ukimwi Road* she criticises the role of NGOs in Sub-Saharan Africa. I've seen the capital projects rising like Ozymandias, never finished, or finished and never used, or used and then abandoned when the machines stopped working and there were no spare parts. Usually, they weren't finished because the money had vanished into a politician's motor fleet. *Look on my Works, ye Mighty, and despair!/Nothing beside remains.* 'My view,' Murphy once revealed in an interview, 'is that I have these things I want to say and I don't really care if it spoils a pure travel book.' NATO, the IMF, the World Bank, the WTO – Murphy was against them all, because (among other reasons) she could see they had deviated from the noble goals to which they had once aspired. I felt the same about the UN. El Salvador was a member state in 1982 when the army and ORDEN, a state-sponsored vigilante force, murdered 5,840 Salvadorean civilians. The two institutions also 'disappeared' (it became a transitive verb in the second half of the twentieth century) over a thousand civilians. And human rights groups consider 1982 a good year in El Salvador. 'USAID is a long pipe,' a Vietnamese told Martha Gellhorn, 'with many holes in it. Only a few drops reach the peasants.' At the time Gellhorn was filing from Vietnam, USAID channelled many millions of relief dollars through various ministries in Hanoi. 'If the Americans want people to get rice,' a Catholic priest told her, 'they must give it out themselves.' This does not seem hard to grasp. Like Murphy, I have heard well-meaning experts point out the fallibilities of aid the world over. I once sat in a dark bar in southern Albania with a Scandinavian NGO man who had been working for years to alleviate child poverty in Albania, a land at that time emerging from the grip of one of the most deranged dictators ever to have lived, and that is a savagely contested field. Before committing, donors wanted promises from the government. 'If I give you something,' said the Scandinavian, simulating the gesture with his hands in the gloom, 'and insist on something back in exchange – that's not aid.' *The lone and level sands stretch far away.* In power relations negotiated through giving I instinctively perceived a macro version of the economic inequality navigated in individual lives and personal relations – the stuff of all writing.

In Malawi in 2006 I saw gleams of hope in microcredit. In the model I covered for the *Telegraph* magazine women received from £15 to £180 in kwacha in a four-month cycle, as well as, crucially, ongoing support. With the loan, I wrote, 'They set up in a variety of fields, from sewing to market trading, and end up with a steady income from a sustainable business.' I sat in on a borrowers' meeting in Sasani and watched a loan officer coaching twelve women in the newly formed Sitigonja group (the name means 'we will not give up' in Chichewa). Twenty-four-year-old Loveness Banda, a microloan beneficiary, showed me her tea shop, a windowless brick room dispensing rice and buns under a handwritten sign announcing, NGONGOLE MAWA – EATING ON CREDIT TOMORROW (meaning, pay up). Banda had dollied the place up by stringing lines of colourful sweet wrappers twisted into butterflies round the walls. 'One day a week is buying day,' she explained when I asked about her routine. 'It's a four-hour walk each way to catch a minibus to the town market, where I pick up supplies.' She had already expanded the business to include an okra stall. But in the years that have elapsed since I was in Malawi and, later, Mozambique, I began to wonder about the microfinance model and its long-term benefits. Feminist lawyer Rafia Zakaria's analysis argues that microloans 'validate the colonial thesis that all reform comes from the west', but it was Bangladeshi economist and social entrepreneur Muhammad Yunus who pioneered microfinance in the seventies. Contrary to conventional financial wisdom, the poor had turned out to be reliable borrowers, the philosophy of aid was turned on its head and Yunus converted his programme into a bank he called Grameen. I had seen branches tucked in the Chittagong Hills close to the border between Bangladesh and Myanmar, catering to, and managing, the needs of illiterate women with no other access to credit. Yunus had fulminated against the failure of the trickle-down model, and, like Gellhorn, had seen money go to the wrong places. The year I visited Malawi, he won the Nobel

Peace Prize. Still I wondered. Did microcredit represent the kind of atomised, consumer-driven empowerment I mentioned in the last chapter rather than a broader political empowerment and structural shift? I feared *Cosmopolitan* Lilongwe was on the cards. Then again, many would say, *Why not?* Here's the only thing that rings true. When it comes to the baked-in inequalities of women's lives, large-scale international charity is failing. As for Murphy, while I was editing this book she died in her sleep in Lismore, County Waterford, aged ninety.

I was able to drop the baby series, which *The Sunday Times* travel editor hated as much as I did, and picked up some proper commissions. When I met the aid worker in Albania the eighteen-month-old Wilf and I had gone on the road for the *New York Times Magazine's The Sophisticated Traveler.* He was blond, as I have said, and people patted his head. Somewhere in the south we were on the beach and I noted that other mothers covered the torsos of their toddlers, so I did the same. Perhaps not as sophisticated as my editor had in mind. It was hard, in 1998, to get any kind of purchase on an emerging Muslim consciousness only thirteen years after the death of Hoxha and six since communism collapsed (that year, 1992, there were fifty cars in Tirana). Women wore short skirts in the cities and people were ahead of their time in the statue-toppling game: in Skanderbeg Square the plinth from which a seventeen-metre bronze Hoxha once surveyed his people stood empty. In the south I felt at home as the landscape looked so much like Greece. How could it not? Greece was inches away. Now, six years on, the country enjoyed the highest pro rata Mercedes ownership in the world. Albanian gangs stole to order from Germany, and red-and-white Bayern Munich pennants still flapped from rear-view mirrors. I paid extra, when hiring a car and driver, for a man who had a driver's licence. Rich Albanians drove their stolen vehicles over the border into Greece every day to go shopping. Frontier guards had recently stopped the prime minister after identifying his official car as a stolen one. We were staying in Dhërmi in what in Greece would be called a taverna, and I had made the owner understand that if it were possible we would like fish for tea. I had seen his small boat on the sand. My son and I were swimming in the bay at sunset and the man puttered out. Pleased with myself, I began acting out a fishing game with the boy Wilf. A minute or two later a small explosion below the surface a few feet away sent a plume of water up like a geyser, and a shower of small fish down on our heads. The taverna owner had dynamited dinner from the Ionian Sea.

Denys Finch Hatton's safaris took many months and involved hundreds of Kikuyu bearers who salted hides and carried canvas champagne coolers on their heads. The bald wanderer guided the Prince of Wales through the East African bush during his 1928 royal tour. Slight as a jockey, with fair hair, the prince, known to his family as David, was thirty-four. Like later incumbents, he was not cut out for the role of heir and in Africa had eyes only for bipeds. A month or two into the safari, runners brought panicky telegrams to camp from the aldermanic Stanley Baldwin urging immediate return, as George V had fallen ill. The royal's assistant private secretary, Captain (later Sir) Alan 'Tommy' Lascelles, reported in his journal at Dodoma, 'He [the prince] looked at me, went out without a word, and spent the remainder of the evening in the successful seduction of a Mrs Barnes, wife of the local commissioner. He told me so himself the next morning.'

I enjoyed the safaris I took with the children when I pursued both Finch Hatton and the non-aldermanic prince, soon to be Edward VIII, through thorn bushes. Once, on assignment years before, I went on an organised walking safari through Zambia's Luangwa Valley. But I am not sure I'd ever do one again. Without a biographical purpose, the modern safari attracts me even less than the modern cruise. Something about being driven to an animal and told to look at it. Is that really very different from a zoo? I went to Rwanda recently and didn't bother with the gorillas. I liked what Noo Saro-Wiwa had to say about safari in Nigeria in her travel book *Looking for Transwonderland* (2012). Gashaka-Gumti National Park has some of the most biodiverse land on the planet. Saro-Wiwa warns that a lack of infrastructure means the visitor on safari needs her own tent, vehicle and warden. Sounds good, if you've been force-fed the Big Five in the crowded lodges of Kenya and Tanzania. Through careful dissection, Saro-Wiwa shows that a function of travel, and one more important than ever, is not so much taking pictures of lions but uncovering layers of history we don't know. You can read the novels of Enugu-born Chimamanda Adichie for a flavour of twentieth- and twenty-first-century Nigeria, and for affecting stories of the tragedy of every human heart.[1] But what about travel writing? It has to record the soft tissue of history: the perishable bit. Saro-Wiwa, born in Port Harcourt, 'a tense oil city' in the south of Nigeria, and raised in Surrey, conjures the Asebu people who migrated from the desert north in the eleventh century and left a thousand soapstone figurines in Esie. Each is unique, with an elaborate hairstyle and individual pose: one figure covers her mouth in exclamation. Barely two hundred visitors a month view these extraordinary sculptures. As Saro-Wiwa comments, 'even the most illustrious parts of Nigerian history seem to be relentlessly buried under time and indifference'. The Saro-Wiwas are Ogoni, a delta people. The author's father Ken campaigned against the depredations of Shell, which has been drilling since oil sprang from Nigerian rock in the fifties. Lagos hanged Ken in

---

1. And in a long essay, 'We Should All Be Feminists' (2014) published as a short book, Adichie wrote, 'Some people ask, "Why the word feminist? Why not just say you are a believer in human rights, or something like that?" Because that would be dishonest. Feminism is, of course, part of human rights in general – but to choose to use the vague expression human rights is to deny the specific and particular problem of gender. It would be a way of pretending that it was not women who have, for centuries, been excluded. It would be a way of denying that the problem of gender targets women.'

1995. *Transwonderland* is a love song in a minor key, to him, and to a country. Noo (pronounced *gnaw*) wrote guidebooks to other West African nations and wondered why Nigeria, 'stretching from the tropical rainforests of the Atlantic coast to the fringes of the Sahara', rarely features on a tourist itinerary. So she went home as a tourist. In the middle belt of Nigeria, where Islamic north meets Christian south, Saro-Wiwa describes jagged 'wild west' scenery filled with hanging rock formations close to Jos, the former tin-mining town on a mosquito-free plateau; the freshness of those high altitudes made Jos her favourite city on childhood pilgrimages. Nok people settled the acidic soil of Jos and founded a civilisation which flourished in parallel with the early Mayan. The author discovers Sukur, a stone-age mountain kingdom in the Mandara Mountains in the northeast. The pages of her book do not indulge in idealism. In the half-century since independence, Saro-Wiwa says 'Nigeria has leapt from one kleptocracy to the next.' Even in the post-*Transwonderland* decade the Nigerian regime has broken its own records in systemic failure. It is an 'unglamorous' destination, corruption 'as inevitable as death'. Saro-Wiwa acknowledges that even the Nigeria her father campaigned against was 'less damaged' than the one the visitor sees today. The book talks of 'tourist brochures', but I've never seen one. Nigeria is a place where 'people and cultures are constantly shifting, disappearing, buried beneath the sands of time and governmental indifference to history'. My point is that we go to the same places all the time, and perhaps there is another way. In Nigeria's 'tourist capital', Calabar, in the estuarine Cross River State, the slave trade, according to Saro-Wiwa, 'transformed the Efiks from fishermen into Anglophile traders'. In today's climate of cultural reckoning the Nigerian account of slavery and its legacy could do with more exposure. Reflecting on the 'harrowing' and frank display at a waxwork slave museum in Calabar, the author writes, 'How I wished we held the cultural monopoly on the portrayal of African history.'

Transwonderland represents an expression of *Heimat* – an untranslatable German word meaning one's native land or region but, crucially, carrying a strong emotional charge. The Welsh *hiraeth* conveys a similar longing for the landscape of the homeland. I envy writers like Saro-Wiwa who can access *Heimat*. I have never experienced it. Dervla Murphy expresses it in the autobiographical *Wheels within Wheels*, her best book. She defines it as 'an element of belonging', continuing, 'There is a sense in which country folk, however impoverished, own their birthplace and all the land around it that can be covered in a long day's tramp – the natural, immemorial link to the territory of a human being.' City-dwellers are exiles from *Heimat*: they are the Dispossessed. (Though Murphy loved pedalling through Dublin among the 'ghosts' of her ancestors.) Money and worldly goods, linked to the city, stand as the opposite of *Heimat*. 'I had been brought up to understand,' Murphy writes, 'that material possessions and physical comfort should never be confused with success, achievement and security.' I had been brought up to understand the opposite. Material possessions *were* success. What else was there to do but *get stuff*? It was though important to be bitter about people who did have stuff. Because that was the stuff we didn't have. Even figures of authority like the family doctor incited bitterness. His wife, who opened the door to us at the King's Drive surgery, was known as the Dragon. Healthcare professionals and the rest of the ruling elite were the Other. If it had been simple class war I would have welcomed it. But those below us in the system were Other as well. Benefit fraud did not feature on the menu of resentments then, but these people, presumably genuinely poor, were painted as malingerers living in sinks of iniquity around Fulwell and Knowle West. Had resentment been based on kinship solidarity perhaps I wouldn't have minded so much either; I would be able to cycle through Bristol among friendly ghosts of my own. But my mother's family didn't like my father's family because they were more common than us, a characteristic manifest in their excessive consumption of sugar. ('She uses a bag a week! I use a bag a month!') Almost everybody, it turned out, fell into the Other category. There were no Damascene moments like the birthday gifts Stark and Murphy describe. But I slowly learned that everything in the mental landscape of the world in which I grew up was wrong. So perhaps *Heimat* is a lot to expect.

At the joint event I mentioned at Somerset House, Murphy was talking about her two books on Palestine. One, *Between River and Sea* (2015), came in at almost five hundred pages. Her wish to assist the Palestinian cause was so sincere that Murphy demoted the importance of literary merit. There was no time for showboating when lives were involved. Murphy's priorities were the right ones. I almost envied her moral compass (which made me think of Alan Bennett writing that he felt a chump for envying the picturesque setting of his friend Russell Harty's grave). It must have been obvious to everyone in Somerset House's Courtauld Theatre that Murphy was a thousand times the woman I was. But that didn't stop her telephoning me from Lismore a week later to tell me she had just that minute finished my latest book and how very much she enjoyed it.

I remain uneasy about the association of womanhood with motherhood. A childless woman is not less of a woman than the other kind. To a certain extent she is more of one, as she is not conforming to Judith Butler's gender norms, or norms belonging to anyone else. I hated the fact that some people saw me as a fulfilled woman because I had children; I mean, they thought I was more fulfilled than I had been before. When I hear of young women getting married in Britain today changing their surname to that of their husband I want to weep. When my feminist conscience dawned in the seventies, mature second-wavers were busy rethinking everyday patterns of women's lives in the departments of housework and childrearing. I caught the end of this in the consciousness-raising groups on Oxford's Cowley Road. I felt that those of us who followed the pioneers had dropped the baton.

What, on the other hand, were the advantages conferred by children in my line of work? I already knew that I had a better chance than my male peers when it came to access, because, consciously or subliminally, people were less afraid a woman might do harm. I had seen this many times over, and was not ashamed to facilitate the process with a smile. Once, in communist Warsaw, I had actually cried in the state accommodation office. A black-swathed trout behind the desk said it was not possible for a single person to take a double room. It was night, and only doubles were available. In an account I wrote of the episode I cried for effect, but I suspect now that the tears were genuine. At any rate, crying failed. (Something else did not. You can read about it in *Access All Areas: Selected Writings 1990–2010*.) I note from the accounts of male peers that they do actually always get where they want without children, so perhaps the female advantage I cite is a delusion.

A child constitutes a double advantage on the road. Not only are people not afraid of you, they actively want to protect you – they give you tin foil. This isn't a delusion. I was in South Africa once, hiking in the Klein Drakensberg with a baby strapped to my chest. His father was with us, so I was not the vulnerable lone female. We were fifty kilometres or so from Hoedspruit, where the mountains are among the most magnificent on earth. I experienced in front of that vista as many have before me the monumental lifting of the spirits that in the human realm only a genius like Schubert can confer. It was a perfect temperature, as it always seems to be in South Africa. A gleaming white Audi purred to a stop alongside us and a meaty white man rolled down the driver's window. I could tell immediately that he was of good intent. 'Need help?' he asked. 'No thanks,' said the baby's dad. 'We're enjoying a walk.' Silence filled the warm air. Another meaty man occupied the passenger seat, and in the back two thin women with custard-coloured hair looked at us. They wore an arsenal of rocks. 'This is not hiking country,' said the man. 'You should return to your vehicle. Those people in the township over there,' he gestured to the north – 'they're not very hospitable.'

I've said it's not hard to take them along when they are babies, if you can face the practicalities. After that you start to see them learning with you – about mosquitoes, about angelfish, about the Other. It was not difficult either during that secondary phase to pick trips they would enjoy: what small boy doesn't love wild camping and its attendant pleasure of not washing? We did that in the Caucasus, in the green valleys that look south to Mount Elbrus where Noah's dove landed. When choices weren't straight-forward I learned to compromise: the youngest and I toured the Boca stadium in Buenos Aires in return for half a day trailing round markets. When their father was with us in Cuba we cut a four-way deal. Five days looking at crumbly architecture in Havana, five days at a coastal resort. This was unexpectedly successful, in that they enjoyed Havana and I loved the resort. The boys played football with Cuban children in the backstreets of the capital using balls made of taped-up newspaper, and went crazy for tank-sized vintage cars: Chevvies, Plymouths, Cadillacs, Dodges – we even spotted a mauve Pontiac, with fins. Their father, a pre-revolutionary North American relic himself, took to shouting '1958 Buick!' or whatever, as every fresh model clunked into view. The following week I found myself in para-dise in Paradisus. This was a resort on the Hicacos peninsula in Matanzas province. (When both sugar and Soviet support failed to keep Cuba afloat, the government decreed tourism the big hard-currency earner, and invested heavily in all-inclusives around Varadero.) The boys were old enough to run around with-out me – they cage you in in these places to avoid contact with anyone who isn't like you. Wilf and Reg engaged in water sports all day, and in between swam up to pool bars with underwater

stools and asked for 'non-alcoholic piña coladas', a phrase now
deployed in our house to indicate a redundant adjective.

Some deals I cut with myself. I had to. In BA I spent half a day
with a writer who took me on a tour of Borges' city. A small boy
would have fainted clean away with the tiresomeness of it all. So
I arranged for him (it was the younger one) to watch children's
films he picked himself at our faux belle époque 1950s boutique
hotel on Rodriguez Peña – I was on assignment – a place so luxe
that it did not display a sign outside, and discreet in the way only
a South American top-end establishment can be. Next day we
headed up into the puna, where there was no electricity. It made
me feel better about the hours of American crap, and that was
the important thing.

They both went through the list stage, when they wanted all
the information in the world delivered to them in a list. So I
would wake up to, 'Mum, who are the top ten goalkeepers ever to
have lived?' and be kept awake by, 'Mum, what are the ten most
common amphibians on the African continent?' Children can
break the ice in groups with their questions, and there is usually
someone who wants to show off, so I became adroit at steering
enquiries in a public direction. But I soon learned to check that
those who like me are bored to tears by other people's children
could escape. For example, I didn't let a list proliferate in a safari
minibus unless I was at the end of my tether. But a campfire was
all right, especially if I cunningly herded the questions into the
wildlife category, creating the general impression that the conversation was indigenous to the scene. I came to see this technique
as an evolutionary device favouring the travelling mother, one
that ensured she didn't perish of ennui but stayed alive until the
next day to nourish her young.

It was tricky, as they grew, to know what they were ready for.
In Yunnan, our loveable guide Apu, who otherwise did nothing
but get us lost, walked us round markets to buy supplies before
we trekked to the Sichuan border. The boys loved chasing a

chicken they had picked for the pot, but when Apu held a bunny up by its ears, they started crying. They found their own way of getting on with China. In Baoshan they peered into open windows where a woman was washing up in a tin basin or weaving on a loom, something that as an adult I could never do on the grounds that it's nosey. People invited them in, and they returned chattering with news. They learned a couple of dozen words of both Mandarin and Naxi. People involved them. At a noodle shop in Lige on Yunnan's Lugu Lake, Reg performed with the resident drummer, and when we went for a walk before breakfast one day, two boys leading water buffalo gave Reg and Wilf a ride. Once, they started fooling around with a ball in a courtyard of the Labrang Monastery on the Tibetan plateau, among the most important Buddhist centres in the world. Within minutes, young monks fluttered out like starlings, hitching up crimson robes and racing for a touch.

They became disrespectful towards me. We were in California once, in the bedroom of a motel overlooking a car park, and I had agreed to do a live television interview for a UK evening news programme: they wanted a talking head to voice an opinion on the fury foaming over little Britain about the phasing out of the old blue passports. I had read enough nationalistic blunt-headedness on this topic, and was happy to oblige. The mood in the motel room turned hysterical during set-up, when, as the video refused to stream, the tech team at Television Centre suggested we might have to record with audio only. 'That would be fine,' said the presenter, thinking he was muted. 'She's not exactly Taylor Swift.' The video popped to life, but Junior considered this the funniest thing said by anyone ever and I had to expel him from the room for excessive noise-making. Now fully engaged in the spirit of the occasion, he stood in the car park, waiting till he could see my jaws moving, then turned round and dropped his shorts.

I had started research on the American book. It was about

middle-aged women writers who reinvented themselves by travelling in the nineteenth century. My boys loved our trips to the US, especially the two house swaps we did back-to-back in California when I was following Jane Austen's niece. I could leave them alone for the day now if I had to go off to some boring (to them) place. Four of my six girls had children of their own, and their challenges put mine in perspective. Rebecca Burlend represented the pioneers, among the most doughty female travellers ever to have lived. Their voices are rarely heard, as they were too busy cutting corn with blunt sickles and delivering their own babies. The God-fearing Burlend had never strayed more than ten miles from her Yorkshire village when she left England in 1831 with a husband, five children and £100 in cash, bound for a lonely bit of sod in western Illinois. When the bedraggled group got off a steam packet at Phillip's Ferry, Rebecca reported years later in a dictated memoir, 'My husband and I looked at each other till we burst into tears, and our children observing our disquietude began to cry bitterly. Is this America, thought I?' The Burlends were among the first to break the soil of the Old Northwest, and Rebecca brought in the harvest while breast milk leaked through her calico dress.

Julia Kristeva in 'Motherhood Today' (not a magazine in WHSmith but an essay serious in the way known only to a French intellectual, apologies, Intellectual) points out that feminine fertility, and pregnancy and motherhood in general, 'serve as a sanctuary for the sacred ... Today motherhood is imbued with what has survived of *religious feeling*.' A friend of mine, a fiction writer, told me that a taxi driver had asked her what she did for a living. Too whacked to get into it, she said, 'I'm a wife and mother.' He was pleased with this answer: 'Nothing finer,' he replied, and meant it. Deborah Levy acknowledges the discomfort some of us feel. In *Things I Don't Want to Know*, written in 2013 in response to George Orwell's 1946 essay, 'Why I Write', Levy reports from the school gate:

> Now that we were mothers we were all shadows of our former selves, chased by the women we used to be before we had children. We really didn't know what to do with her, this fierce independent young woman who followed us about, shouting and pointing the finger while we wheeled our buggies in the English rain.

Levy said that she and the other mothers were not so much grown up as grown down. I had the advantage at least of having my children at a venerable age – I had the second at forty-one (the NHS uses the clinical term 'geriatric mother'). I ran a campaign at school demanding a handicap in the sports day mothers' race on the grounds that there is no athletics contest in the world in which one competes against runners twenty years one's junior (it failed, and I came last). In *Things* Levy goes on to say that motherhood is 'an institution fathered by masculine consciousness', and my friend's taxi driver bears the statement out. So too do my experiences on the road. Having a child in tow satisfies something nebulous in people's minds: with a child I am in the right place; unthreatening not just in the sense that I am unlikely to be about to stage a robbery, but in a general sense of maintaining the natural order of things. I did not relish performing that role. I'm not sure if I was Levy's 'shadow of my former self', but I was better off, on the road, on my own. One spots in the taxi driver's reverence a throwback to the nineteenth-century American cult of true womanhood – stay on the pedestal, dear. Sojourner Truth famously showed the nastiness and hypocrisy of that notion, you'd have hoped once and for all, but we know that to be wishful thinking. Speaking at the Ohio Women's Rights Convention in 1851, Truth, a former slave, reportedly asked the assembly of white men and women where her pedestal was. Over the objections of the white women's-rights advocates who sought to silence her that day, Truth spoke of the brutality she had endured and wondered aloud why she didn't receive preferential treatment as a member of the fairer sex, famously asking, 'Ain't I a woman?'

(Oh but did she ask it? Recent scholarship casts doubt on the much-quoted quote. Professor Nell Irvin Painter convincingly argues that a white woman came up with a variation on the phrase (*Ar'n't I a woman?*) while reporting the Akron speech, and that subsequent feminists have blithely 'reworked the line into a supposedly more authentic dialect'. Truth spoke perfectly standard New York English. The celebrated line has in any event taken on a life of its own 'as a synecdoche for what we now term "intersectionality". The false quote flattens Truth into little more than a magical Negro saviour of white women.')

Today the subject of motherhood's tendency to grow us down looks like self-indulgence. Should women work? Books with titles such as *Loving and Loathing Our Inner Housewife* – I can't be doing with it. I return often, on the other hand, to a take on the subject by Deesha Philyaw which appeared on Bitchmedia:

> Low-income and working-class women, Black women, and other women of color don't see their mothering experiences and concerns reflected in the mommy media machine, and we get the cultural message loud and clear: Affluent white women are the only mothers who really matter. Further, media over-exposure of these women bolsters the perception of them as self-absorbed brewers of tempests in teapots.

Mummy memoirs on the whole are a crashing bore, though I like some of the titles (*The Three-Martini Playdate* covers a multitude of synecdoches). In the end I pursued feminist theory so far on the motherhood topic that I wanted to shoot myself, so I'll leave it there.

5

Not Even Past

North America

In 1979, the summer Hurricane David made landfall and I turned eighteen, I toured the US on a Greyhound bus. Thirty-two years later limos whizzed me to speaking and signing gigs on a winter publicity tour through the same country. I suppose the key events of my life occurred between those two journeys. I loved the United States the same from ripped plastic seat to luxury leather recliner. Americans try so hard to be *helpful*. On a magazine assignment in a town that might have been called Coma, I asked the motel owner if he could possibly provide a crib (cot), as I had forgotten to book one in advance. The motel didn't have such a thing, but the man kindly said he'd see what he could do. An hour later he wheeled in a shopping trolley. As Walt Whitman wrote in the preface to *Leaves of Grass* (first published in 1855), 'The United States themselves are essentially the greatest poem.'

Places that have not quite melted in the pot – those are the ones I like in the US. Milieux alien to a European; vestiges of pocket-size Americas that once existed from sea to shining sea. Backwaters, usually. I once drove 150 miles from New Orleans to the swamps of southwest Louisiana to find out what 'Cajun' means in modern America, beyond a coating of chilli seasoning.

A companion, Marcel, had offered to show me round the old sugar plantations of New Iberia and Lafayette County. Marcel had left home for The Big Easy twenty years earlier, and made a good living as a software engineer. This was 2010, and he was mad about the monster oil spill that had erupted over the Gulf Coast three months previously. Everyone in New Orleans was mad. A radio station was advertising a crawfish cook-off with a stall at which one could smash a plastic BP logo with a mallet. Oil was flowing east from *Deepwater Horizon* so far, largely because of tidal patterns, leaving Cajun bayous unmolested. But that was so far. Self-absorbed oil executives were still treating the disaster as a tempest in a teapot.

Across the cane fields, mist rose from Bayou Teche, veiling a sugarhouse weathered black. Marcel, a true-born Cajun, looked out towards the porch of a decaying planter's mansion at a Mississippi kite describing loops in the stifled air. 'Welcome to America's greatest backwater,' he said. Highway 182 crept west across the lowlands and their farm supply outlets, piercing acres of cane. Gun racks rattled on pickups. We stopped at St Martinville, on one of the Hurricane Evacuation Routes that crosshatch Louisiana. The names on the mailboxes were French: Boudreaux, Thevenet, Broussard, Hebert. Spanish moss bearded the oaks and so did tightly curled fronds of resurrection fern, an epiphyte that comes back to life in the rain. At the end of Main Street, stragglers hurried to catch 9.30 mass at St Martin de Tours, for two and a half centuries the mother church of all Cajuns, including atheists. French-speaking Cajuns came to Louisiana as pioneers, backwoodsmen and exiles twice over. Their forebears, poor Poitou farmers, had migrated to what is now Maritime Canada in 1604. France wanted fur and cod, and the settlement, which they named Acadia, flourished in a temperate climate more continental than maritime. But when the British acquired the region in 1713 and rechristened it Nova Scotia, Acadians refused to swear an oath of allegiance to the

Crown, and in 1755, after years of harassment, George II ordered his colonial administrators to throw them out. At least half the Acadian population died during what they still call *Le Grand Dérangement*, either from smallpox or typhus or starvation. But from 1764, three thousand Acadians made their home in the hot and barely populated prairies and wetlands of southwest Louisiana. The clan had adapted once; now they would adapt again. A new Acadia rose from the jungle like resurrection fern, and the name of its hardworking people evolved from Acadian, to Cadian, to Cajun. They knew how to plant, fish and hunt, and the land, though inhospitable, was fertile. In time, by dint of hard work and endurance, Cajuns became the first group of European colonists to acquire and retain a distinctive North American identity.

In the first decades of the twentieth century, 'Cajun' became a term of abuse. It meant white trash. But eventually the zeitgeist changed, as it always does, and ethnic revival stoked interest in things Cajun. Local businesses rushed to incorporate the word 'Cajun' into their names, the rest of America tuned in to Swamp Pop and the state legislature in Baton Rouge officially designated twenty-two parishes Acadiana. (More recently, detective writers have pioneered bayou noir.) For more than two centuries Cajuns have walked a tightrope between assimilation and independence. As late as the seventies, the state mandated that French had to be taught in primary schools, distributing posters exhorting parents, *Parlez français avec vos enfants à la maison*. At that time, attempts to preserve Cajun culture focused on language. Now, nobody speaks French. Language preservationists battled not only television and English-speaking incomers, but also a stigma French has always held for Cajuns – one associated with ignorant swamp-dwellers gabbing with comical accents. But I fancied that a lilt of old Poitou lingered when a waitress took our order.

The Cajun community of Louisiana today, numbering several hundred thousand, remains relatively homogeneous and rooted in the region. (Marcel was the only one of his extended family who had left, and he had not gone far.) In the American psyche, over two generations Cajuns have come to exemplify a brand of noble endurance. Their survival in a mysterious region where alligators rear from primordial swamp evokes a tribal capacity to win through, an increasingly attractive model in our gruesomely individualistic age. Still Cajun in his heart, Marcel liked to boast about food. At his family home, his mother ladled out crawfish *etouffée*, a hot gumbo of bony crawfish tails and unctuous white gravy. It was hard to get away from crawfish. (Britons know them as crayfish.) The legend goes that when King George expelled the Acadians from Nova Scotia, the lobsters followed. During the swim down the coast of North America, the crustaceans grew thinner and thinner, ending up as the crawfish endemic to the swamps and bayous. There's not much to a crawfish, which is perhaps why spice, as everyone knows, is the determining factor in Cajun cuisine. Marcel had plenty of complaints about the bland meals he had been served on a recent trip to Paris. (As I was to learn in Quebec, North American Francophones complain keenly about all things *parisiens*, including the puny crawfish known as *écrevisses*.) In fact, Cajun cooking blends French elements (a ubiquitous brown roux came via Nova Scotia) with aspects of West African cuisine (okra arrived with slaves from what is now Mali) and a Spanish vibe (the region was once a Spanish colony). The first real Americans contributed spicy filé powder ground from dried sassafras leaves. If truth be told, and Marcel wasn't going to be the one to tell it, you can eat better in the restaurants of New Orleans' Garden District than you can in the Atchafalaya Basin. But I loved the Cajun lunch counters strung out along country roads. Waitresses called me baby and food came in paper bags rather than on plates. *Boudin blanc* – white blood pudding – had to be squeezed directly into the mouth from its pig-gut stocking, a complex sucking and gobbling manoeuvre that cut my appetite. Like the tastiest of its pork crackling, bayou country is an underbelly, south even of the Bible Belt, which by common consent starts at I-10. Left to itself and isolated, the indigenous ecosystem has flourished more or less unmolested, like the Cajuns. From a shadowed glade on Bayou Petite Anse I watched snowy egrets and roseate spoonbills skimming the surface of a two-hundred-foot pool, and a tiny tricoloured heron fed under a sweet gum. Nothing else moved. The landscape, infused with the blue light of the bayous, could not have been mistaken for any other place in North America. I never saw a gator, but I did see Gator Autos on the outskirts of Lafayette. Marcel had tried to evoke the amalgam of America and seventeenth-century France that has shaped that enigmatic corner of rural Louisiana. True Cajun culture is not really about music and food, though both are a cause for celebration. It is about the weight of history, and something indefinably other that has survived the centuries along with folk memories of a good French roux and a clannish determination to keep going, whatever the next hurricane or government might bring. The past pressed on in the present despite everything. We drove back to New Orleans, ramping up the air con. Just before we left the rural roads to join an interstate, Marcel pointed at a sign raised high on the bank of a drainage canal. 'Look at that!' he said, as if he had found what he was looking for. 'See the K – as in Kool Aid?' I looked towards the bayou, across a patch of swamp toothed with cypress stumps. In the foreground, the sign advertised KAJUN DONUTS.

Five years later I encountered another group of outsiders in Louisiana. They were inside. I was temporarily inside too, but outside: in a ten-thousand-seater stadium in the Tunica Hills, where bulls of Minotaurean proportions stamped up clouds of dirt. Late in the afternoon the bulls retired and the crowd filed out while a pair of central-casting clowns entertained the children. But most people could not leave. They were prisoners.

The Louisiana State Penitentiary, known as Angola, is the

largest maximum-security jail in America. Since 1965, it has hosted a rodeo six times a year. Three-quarters of the 6,300 Angola inmates are African American, the majority lifers, ineligible for parole. A hundred or so are on death row. The rodeo takes place every Sunday in October, and over one weekend in April; the public can purchase tickets for $15. It's a slow three-hour journey from New Orleans. The prison sprawls over eighteen thousand acres of former plantation land. One of the cotton estates there was called Angola, after the place its slaves came from, and the name stuck. The site opened as a prison in 1901. Bordered on three sides by the Mississippi and on the fourth by miles of inhospitable woodland, it forms a kind of island, and, once you put up fences, a natural jail.

An MC in a Stetson choreographed the rodeo from a piebald horse. He shouted into a microphone, and waved a star-spangled banner on a long pole, an activity that made him look like a knight. Competitors – all inmates – wore striped shirts, jeans and spurs; other prisoners sported an easily identifiable white shirt and sat in caged pens in a section of the arena. Special performances alternated with horse races and bull running. They called one contest Convict Poker. Four men attempted to remain seated at a card table while a bull charged. They lasted a minute before sprawling in the dust. At least they wore helmets. I expected an ambulance every second. Before and after the rodeo, the prison puts on an open-air arts and crafts fair. A display of tropical hardwood and oak goods filled at least an acre, as well as oil paintings, jewellery and garden furniture. I was interested in a pot stand – a platter on legs upon which one places hot dishes at table. A white man of about seventy had spread a selection on his stall. He quietly explained how he had made each one, and which woods worked most easily. When I chose a handsome piece fashioned from triangular segments of oak, each with the grain going in a different direction, the man gave me a chit, which I took to a payment booth. Then I returned to the carpenter and collected my pot stand. 'I hope you enjoy it,' he said quietly. I could have wept. *What did you do?* Rodeo privileges are earned. Only men with a solid record are allowed to sell craft items or work on food stands. Others with just a few years of good behaviour are contained, before the actual rodeo, in a cage about the size of a tennis court. Some of this group loitered near the perimeter fence. One man heard my accent and told me he had visited London on a trip with his wife's firm, and had liked Tower Bridge. Those who had not accrued good behaviour years remained in their cells.

Warden Burl Cain had a robust reputation both in Louisiana and throughout the US penitential system for reducing the violence which dogged Angola for decades. (His Biblical namesake was the first murderer. I don't know if this counts as irony.) A native of the Pelican State and a devout Christian of the evangelical Baptist variety, Cain encouraged spiritual study. 'Our greatest challenge,' he has repeatedly said in interviews, 'is to give hope where there is hopelessness.' He was a Republican who earlier that year had announced his intention to stand as state governor. But he had his detractors. Former prisoner John Thompson, who spent fourteen years on death row before being pardoned, once said that Cain runs Angola 'with a Bible in one hand and a sword in the other. And when the chips are down, he drops the Bible.'[1]

Inmates' family members can purchase rodeo tickets. While perusing craft stands, I saw a prisoner sitting on a plastic chair holding his mother's hand. He can't have been older than twenty-five, but he must have been inside for years already, to be allowed to roam outside the cage. Ninety per cent of prisoners will die in jail. There is a hospice as well as a hospital. If nobody comes to collect the body – as is often the case – prisoners fashion a coffin, and a preacher conducts a funeral. *The Angolite*, the prisoners'

---

1. In 2018, Cain resigned in the wake of allegations of impropriety of which he was subsequently cleared.

newspaper, records an obituary of each man, always ending with
the words, 'Buried with dignity.'

I drove from New Orleans up the levee road to Mississippi,
travelling north to the South. The sky was cloudless. By ten I
had cleared Baton Rouge, and north of Vicksburg picked up
Mississippi 1 at Mayersville. The fields dissolved into pea-green
soup, the spawny surface of the swamp holed with stumps and
sweet bay magnolia. One expected a dinosaur to surge through
the black gum, sending spumy flumes up hundreds of feet. When
the ground solidified, levee and road diverged around the fla-
mingo legs of a water tower, and thousands of acres of corn rolled
out ahead. Densely packed, ten-foot stems shot out leaves broad
as flags. There was something impressive in the plants' implaca-
ble determination to swell in every direction. It did not surprise
me to learn that the Mississippi Delta yields three crops a year.
Three a night would not have surprised me. The Delta is regularly
cited as the distilled essence of the Deep South – the last Dixie
stronghold. Properly the Yazoo–Mississippi Delta, its alluvial
plain extends two hundred miles from Vicksburg to Memphis,
and the host of a local radio show once boasted that the blues
were born at the Tutwiler railhead. Outside Greenville, a funeral
cortege blocked the road and a row of construction workers
stood with hard hats covering their hearts, the air around them
spangled with corn dust. In the brakes, painted houses nestled
among rowan oaks, netted windows black blanks. A chunky
white man piloted a mower across a lawn. The air drooped. The
radio jock had a studio guest who had come on to demonstrate
that raising federal income tax was unbiblical. This was not an
issue discussed on the pages of *The Angolite*.

It doesn't matter how much or how little a reader knows about
an author's life. But isn't it thrilling to see where a writer sat
when he or she conjured characters out of the air? In northern
Mississippi I diverted to Oxford in Lafayette County and William

Faulkner's house: the novelist's central theme – the importance of the fight to be freely oneself – was locked in my black traveller's heart. The flat green land-sea that rolled back from the levee as the Honda barrelled north provided the backdrop for 1939's *The Old Man*, Faulkner's novella set in 1927, the year of the great flood. The Mississippi plays the title role. I learned more about the South from Faulkner than from anyone, or at least about that small area of northeastern Mississippi which inspired him – what he called his *postage-stamp world*. Inside his house, Rowan Oak, sunshine flooded through well-proportioned windows and the silence of the woods came with it, barring the confidential creak of my own footfall. Cedars cast lacy patterns on the kitchen tiles and over the hollowed keys of Faulkner's Underwood Universal Portable. It was such a perfect writer's house that one wondered why Faulkner drank so much. From the cool of his rear study, he ran the domestic world that provided a foil to his creativity. He smoked hams in a brick smokehouse, kept a cow in a timber barn and petted Tempy, his favourite horse, in stables on the west side of his paddock. He had created his own backwater, a pocket-sized America in which men and women grow broad as corn leaves. Walking among the sweet shrub of his maze, Faulkner composed the acceptance speech he made in Stockholm in 1950 when he collected his Nobel Prize. 'I believe,' he said, 'that man will not merely endure, he will prevail.' He had not been to an Angola rodeo, and anyway, the evidence that man as a species will endure is weaker than it was when Faulkner stepped up to the podium. Yet his belief in personal endurance resonates as one sees one's own crises zip by. I had felt that increasingly as the decades passed between my American journeys. Faulkner was obsessed with the truth of history. 'The past is never dead,' he wrote in *Requiem for a Nun* (1951). 'It's not even past.' The Cajun story suggested he was right. But fiction never lies. The novels indicate that Faulkner glimpsed the annihilation to come. It's enough to drive anyone to drink.

Imagine a map in which the world is not divided into nations but apportioned to writers who have imprinted the place so indelibly on the page that it will always belong to them. Rowan Oaks, its cedars and smokehouse, belong to Faulkner, but the Delta belongs to Eudora Welty. It was a place, she wrote, where 'most of the world seemed sky', a landscape of the imagination, like Housman's Shropshire, and Welty's books are gazetteers of the heart. She was born in 1909 in a house her father built in Jackson. I went to pay homage there once. The family's antecedents were Swiss immigrants who cut a path through the Yazoo wilderness in the 1820s. A few years later the retreating Choctaw, loyal custodians of the Delta, ceded their fertile territory to the new American government at the Treaty of Dancing Rabbit Creek, another North American *grand dérangement* and unsung human tragedy. Welty though remembered family trips in an Oakland touring car, careering down a roadless bank 'with daddy simply aiming at the two-plank ferry gangway'. Another time, when she was ten, she and her father, an insurance man, sat on folding chairs at the rail of the open-air observation platform of a train. 'We watched the sparks we made,' she reported, 'fly behind us into the night . . . the sleeping countryside seemed itself to open a way through for our passage, then close again behind us.' The train appears in Welty's best-known book, the 1946 novel *Delta Wedding*, which opens with nine-year-old Laura McRaven riding the Yellow Dog with her ticket in her hat. I took that train once, and I doubt I was the first acolyte to slip my ticket under my cap.

Unlike many writers who return remorselessly to their native soil in their work, Welty never really left Jackson; she said she remained there 'underfoot' her whole life. This again represented *Heimat*, despite the fact that she stayed – a living, faithful *Heimat*. The Delta remains a character as Welty's work moves through the Second War to shattering racial division, agricultural mechanisation and economic hardship. In her first collection of short stories, *A Curtain of Green* (1941), Welty wrote about antebellum homes on a bluff on the Natchez Trace – I knew they were still there as I had driven past them on my way north – and in *The Robber Bridegroom* (1942) she brought back the Natchez Indians whom the French annihilated in 1732: their ghosts still roam Delta forests. She recognised complexity and avoided the romantic legend of the Old South and its platitudes. The reader will find no sentimental regard for the antique order in the pages of *Delta Wedding* (1946), and no judgement of its horrifying racial injustices either. Welty had an ear for the cadences of Mississippi speech, but what she really wanted to write about was 'the voiceless life of the human imagination'. (You might ask what writer doesn't.) A critic describes her genius for conjuring 'the ingrown, post-historic, Coca-Cola-sodden South'. She was an accomplished photographer as well as a writer (she won a Pulitzer and the Presidential Medal of Freedom), shooting mainly with a Rolleiflex. In the Depression she worked as a feature writer for the state branch of the Works Progress Administration, Roosevelt's job-creation scheme, travelling to each of the eighty-two county seats in Mississippi by public bus or behind the wheel of her mother's Chevrolet sedan, observing political rallies, revival meetings and a mule-powered cane-syrup mill. She immortalised the poor tomato farmers around Crystal Springs in Copiah County, and after spotting an open ironing board in a post office turned it into a comic soliloquy in the story 'Why I Live at the P.O.'. Like Burl Cain but with more credibility, Welty found hope in hopelessness.

It was on road trips like that one through Mississippi, and on the much earlier Greyhound journeys, that I glimpsed the underbelly of America. The people on the buses with me – fellow travellers – were from there. I recall waking in a bone-cold Alabama dawn to see a young man asleep in the luggage rack. He had told me he was looking for picking jobs, as he shifted north through the year, following the seasons, and had just finished working the runner peanut diggers in Dothan. On the

2010 motor tour I continued on the back roads to Tennessee. A radio station plugged a Christian Fun Day at Wolf River Mall in Germantown, a glossy southeastern Memphis suburb. At the same time, a purple and yellow hand-painted sign welcomed me to Jonestown in Coahoma County. Lopsided tin-roofed houses on cinder blocks lined the road between litter-strewn vacant lots. The metal double doors of the single grocery store looked as if they had been shot at. Closer inspection revealed that they had. Alongside the doors, half a dozen teenagers squatted among tyres and broken-down white goods. There was none of Faulkner's prevailing spirit here, but plenty of evidence of the failure of Burl Cain's Grand Old Party. Jonestown bumps along at the bottom of every health and education table while topping out in the unemployment and crime stakes. Ninety-three per cent of Jonestown residents are African American. According to government figures, more than 57 per cent of children live below the poverty line – and this on the most fertile agricultural land on the planet.

Nobody treated me as a freak in the South, and I wondered if that meant there was more latitude for individuality. One might expect less accommodation of the lone female, given the region's reputation for being unspecifically unevolved. But I had an open welcome. Notwithstanding my aversion to generalisation, I can say I found fecund wonderment between I-10 and the Mason–Dixon line. Rush Week 1980 at the state university of Alabama in Tuscaloosa was more foreign, though, than anything I had seen (though I had not seen much). I barely needed to consult my notebook to write this, as I remember the scenes so well, and not for the first time while working on this book, I wondered why I had not written about them before. I stayed in student accommodation as a guest of Lynette, a sophomore I had met on a plane, and as 'bid day' approached she walked me round Sorority Row while freshwomen *rushed* to gain membership of a favoured society. On campus we ran into Lynette's boyfriend. They had a date that night, and were heading to a Delta Kappa Epsilon (founded

Yale, 1844) fraternity party, and the boyfriend enquired what beverage Lynette would like him to buy for her. She asked for Diet Dr Pepper. It wasn't quite *Gone with the Wind*, but the lineage was there. At the eight-pillared neo-Palladian Alpha Delta Pi sorority house where Lynette belonged, she pointed out the motto over the door, WE LIVE FOR EACH OTHER. Her mother had been an Alpha Delta Pi girl before her. Six women had founded the sorority's precursor, the Adelphean Society, at Wesleyan Female College in Macon, Georgia, in 1851 – the first 'secret society' for women in America – and after it evolved into Alpha Delta Pi, a branch budded in the Heart of Dixie which grew into a proud member of the Alabama Panhellenic Association. Today students call rush week 'sisterhood round'. I saw no Black faces in 1980. Sorority websites reveal almost none today – perhaps one in a hundred. On the national Alpha Delta Pi website, out of twenty-one Notable Members in the Entertainment and Arts category,

ten are listed for their credentials as beauty queens. Foreign? I think so.[1]

In 2011, obliged to travel to Kentucky for research purposes, I packed up the family and arranged a house swap in a four-thousand-square-foot antebellum brick mansion on the outskirts of Lexington. As it was set in the middle of a horse farm, every window looked onto a thoroughbred mare and her foals grazing on prairie grass and coneflowers. (As part of the exchange, we had membership of our swappees' country club. It was a little – and not disagreeably – like visiting another planet.) The Kentucky River drains the bluegrass wold that makes up the northern part of the state, and between Madison County and Frankfort the river cuts a canyon through the Palisades, a hundred-mile stretch of striated gorges dotted with yellowwood, blue ash and sugar maple. On the *Dixie Belle* sternwheeler, I sat upfront with the skipper, a Captain Birdseye figure with a handlebar moustache and a peaked mariner's cap. He was noting towheads on a chart. Around Shawnee Creek, a motorboat powered between us and the bank, drowning the puttering of the *Dixie Belle* and sending up a shower of great spoonbills. 'Folks from Ohio who can't drive boats,' growled the captain. The old civil war enmity persisted in the border states. Sundays, white folk left rivertown churches in sartorial splendour and processed to McDonald's for lunch. Interstates have allowed Kentucky turnpikes to fossilise. Among white fences, Adirondack chairs and blue spruce, country stores and worn-out truck stops struggle to cling on: a vanishing backwoods America. On route 169 south of Lexington, the Valley View car ferry (free of charge) is one of the few remaining river ferries in the US, operating in a literal backwater. The service started in 1780 when a veteran of the Revolutionary War acquired

1. On 30 October 2021, anti-Semites broke into Tau Kappa Epsilon, a chapter with a significant Jewish membership in DC's George Washington University, and destroyed its replica Torah. They ripped the scroll apart and poured detergent over the fragments. This type of behaviour, appallingly, no longer comes over as 'foreign'.

the surrounding land (this was before Kentucky joined the Union and Spain lost control of the Mississippi). The rudderless, cable-guided ferry holds the record as the commonwealth's oldest continually operating business. It takes three cars at a time, and has flowering window boxes.

Intrigued by Man O' War Boulevard and Man O' War Church of God near our house, I asked the pastor of the latter which name came first. 'Neither one,' said he. 'Both were named after a racehorse.' Early in the morning training sessions at Keeneland Race Park are open to the public. The horses run on a specially developed surface made of golf balls and the outer layer of electric cable, rather like the stuff BP shot into *Deepwater Horizon* in a failed attempt to seal off the oil flow. After watching horses gallop we went to the Track Kitchen next to the drive-through betting parlour for breakfast with the trainers. 'We breed for speed, speed, speed,' said one, making his way through a pile of biscuits fighting for their lives in sausage gravy. 'The horses have knees the size of my twelve-year-old and carry 1,200 pounds – which is why there's no jump racing here.' He paused, as on the other side of the track a jet engine roared and we could not hear ourselves speak. They built Blue Grass Airport right next to Keeneland so sheikhs can walk from their private 747s to see their beauties run.

From the start, I came across unheralded women who had sailed the seas and done extraordinary things that were worth doing – by which I mean they had not climbed the north face with one hand tied behind their back just because it had not been done before. I began actively seeking these shadowy figures to give them their due on the page. They were always there, lurking behind glaciers and dunes. The reader has met some of them already. Since it was looking increasingly unlikely that I would ever do anything worthwhile myself, I started to write their stories down in a more formal way. In Georgia, I had found Fanny Kemble (1809–93). She was the author of an anti-slavery tract so powerful that it incited the government in London to withdraw support for the Confederacy. The Georgia Historical Society had erected a metal sign in the cordgrass stating that Kemble *influenced the outcome of the Civil War*.

I grew up in a slave city. The gigs of my early teenage years – the ones you remember for ever, in my case Nazareth, Slade, Roxy Music – they all took place at the Colston Hall. My mother went to one of the Colston schools. We were knee-deep in Colstons. After school we sometimes walked up Clifton Down past mansions of giddying splendour that we did not then know were built on slaves' backs. Decades on but some years before BLM I had read Kemble's book, and I followed her to the Georgia Sea Islands in a dimple of the South Atlantic Bight. Like the other places I liked most in the United States, the Sea Islands were unique and extreme. They aren't even land really – skeins of oat roots catch and hold mushy blobs like a net, otherwise they would float free in the ocean. High tides sloshing across the basin nourish hundreds of thousands of acres of subtropical salt marsh, and on the mainland the Altamaha River, pouring down from the piedmont, infuses freshwater into the marsh, creating spongy islands called *hammocks* in an inland delta. A thousand years ago, Timucua peoples hunted right whale from the hammocks, living off oysters, shrimp and crab. Conquistadores annihilated the Timucua, the combined efforts of pirates, Yamacraw and Britons disposed of the Spaniards, and Old World planters found that tidewater hammocks made an ideal environment for large-scale rice production. They shipped in West African slaves who spoke the Gullah language and knew how to cultivate *Oryza glaberrima*. On the Greyhound trip, it was the only place I saw evidence of Hurricane David.

Kemble, born into a thespian dynasty in London in 1809, was a Shakespearean leading lady who, having conquered Europe, crossed the Atlantic on one of the first American theatre tours. There she had met and married a front-row beau,

sideburn-sporting Pierce Butler, pro-South radical and scion of
a prominent slave-owning family. In 1838 he took Fanny and
their two children to Georgia's Butler Island. For a generation,
tidewater landowners like the Butlers had reaped bountiful prof-
its from the Sea Islands, and lately rice had overtaken cotton as
the dominant crop. Butler Island, with nine hundred acres under
cultivation, was the engine of the family fortune. In a letter home,
Kemble described the landscape as 'quite the most amphibious
piece of creation . . . 'Tis neither liquid nor solid but a kind of
mud sponge.' As for the human landscape – in the first week,
Kemble met a pregnant woman who had collapsed in the fields,
whereupon the overseer had summoned her husband to string
her up by her hands and thrash her with a hide whip, and when
he had finished, she miscarried. The slaves lived in filth, most
adults were crippled by rheumatism and the older ones crawled
to approach Kemble as their arthritic legs no longer functioned.
Psyche, the slave who assisted the Butler children's Scottish
nanny, revealed that her husband was up for sale, and begged
Kemble to intercede on her behalf. (Between 1790 and 1860 more
than a million slaves were shipped from the Upper South to the
Lower in this domestic trade.) Georgia was howling, but nobody
was listening; William Sherman was still at West Point. Tara was
the plantation to be on, if you were a slave.

By the middle of February, when slaves were burning stubble
and ivory blossom foamed from the narcissus, the Butlers moved
to their cotton estate at the northern tip of St Simons Island, fifteen miles away. The family home there jutted between a creek
and Hampton River, five miles from Frederica, the original forti-
fied settlement. Kemble kept a journal. Her outrage at her family's
treatment of slaves broke the marriage. Butler got everything,
including custody. In September 1845 Kemble retreated home
to London, beaten only in a superficial sense. She was thirty-six.
Sixteen years later, when Georgia followed South Carolina and
seceded from the Union, Kemble was no longer a young woman

casting spells from the stage. She got out her old diary and turned herself into a writer, publishing an account of the scenes she had witnessed in order to influence public opinion against the slave-owning South. It is hard to think of a more vivid portrayal of slavery in the southeastern United States than *Journal of a Residence on a Georgia Plantation*. The stories Kemble told include that of Roswell King Jr, who had occupied the post of resident manager at the Butler plantations for two decades. King had fathered a tribe of mixed-race children. If a woman resisted, he ordered her to be flogged, so she would not resist again. On one occasion, two slaves gave birth to his progeny in the same week. Mrs King appeared a few days after the confinements, personally oversaw the flogging of the new mothers, and ordered the punishment to be repeated daily for a week. At the Butler Island slave hospital, Kemble reported, a four-room structure of whitewashed wood, dying slaves lay in their own excrement, without sheets, medicines, beds or food. 'And here,' Kemble wrote, 'in their hour of sickness and suffering, lay those whose health and strength are spent in unrequited labour for us.' Some, including teenagers, had a disease that rotted their hands and feet. Babies suffered from a particular type of lockjaw. When they died, they were not buried with dignity. Meanwhile barges piled with cotton floated past like ghosts in the current. News of the Siege of Vicksburg broke as copies of Kemble's book reached the binders. *Harper's New Monthly* called it 'the most powerful anti-slavery book yet written' and Henry James said it was 'the most valuable account of impressions begotten of that old southern life which we are apt to see today through a haze of Indian summer'.

The plantations withered without slaves, and the population shifted inland, leaving St Simons to sink back to its un-Jamesian miasma of refracted light and oyster middens. I once rented a car in Savannah and drove across a causeway onto the island at the long, still moment of dusk. The air had a salty tang, and that violet iridescence of light particular to tropical marshes. The yellowgrass, bleak from a distance, close up clicked and whirred with migratory birds and fishy nurseries. St Simons was a low-rise sprawl of boardwalks and simple summer houses. A gated residential development had engulfed the ruins of the house in which Kemble had scratched in her journal. Besides a Pierce Butler Road, the municipality had called one street Roswell King Road. And now his name was painted on a sign, evidence, as if it were required, that words can lie just as effectively as they tell the truth.

On my writers' map of the world, a slim segment of Georgia on the banks of the Chattahoochee 260 miles east of the Sea Islands bears the mark CARSON McCULLERS. The pale-skinned McCullers (1917–67) grew up in Columbus, lived mainly in New York and wrote dazzling stories about Southern freaks. As the reader will have noticed, I like a freak. *The Heart is a Lonely Hunter* made McCullers famous at the age of twenty-three. She married the same violent man twice and a straight narrative has usually framed her lesbian existence. Recently unembargoed tapes of therapy sessions with the woman who became McCullers' lover reveal, the memoirist–biographer Jenn Shapland convincingly argues, 'the only story McCullers ever wrote: a lonely misfit wrestles with her hidden self, unable to articulate her own longings'. She was a traveller motivated not by a desire to see the world or even a need to make sense of it, but by a restlessness with no name. She stayed seven times at Yaddo, the writers' retreat in upstate New York, at times a freak show itself. McCullers' physical afflictions included strokes and temporary blindness, and she also fought that enemy so faithful to writers, the bottle. She shared a rental property in Brooklyn with W. H. Auden (McCullers called him Winston; he stood over the grave at her funeral), Gypsy Rose Lee, Benjamin Britten and others. McCullers in her prose speaks clearly and universally of the human heart, and specifically of the human heart in conflict with itself. In the end, what else is there to write about? When you take the back roads through Georgia the 'two- and three-storey shops and business offices' of McCullers' Main Street still steam quietly in summer sun and the diner at night continues to cast 'a sharp yellow rectangle on the sidewalk'. No vicissitudes have upended those towns like the tumults that have shaped – for example – Key West, once the richest town in America. There are no second acts in the post-industrial mill towns where McCullers' misfits die without dignity. Walking through the streets of Key West in the early months of the millennium, on the other hand, I had a sense of being inside a kaleidoscope. I had driven in over the mighty Overseas Highway. Of the ten thousand islands that curl out between the Atlantic and the Gulf of Mexico, fewer than fifty are inhabited, but the OH strings together the whole pearly set of Florida Keys. This heroic road pursues the arc of islands bridge by bridge (one is seven miles long), finally running out of land ninety miles from Cuba. Key West is the southernmost, a rangy resort town famed for its peculiar blend of seedy tropical chic. Following the pirates, wreckers moved in, plundering vessels that foundered on the west coast's coral reefs. When Cubans poured in seeking refuge, at that time from Spaniards not communists, cigars and sponges supplanted wrecking. By 1880 Key West was the largest settlement in Florida as well as the richest in the land. But by the 1930s, the kaleidoscope had turned.

The island covers eight square miles, and the old town, where tourists congregate, is squeezed into the northwest corner, on the harbour. For many decades Key West has enjoyed a reputation for outré glamour: a haven for artists, cranks, homosexuals and assorted non-conformists in southerly flight from the nine-to-five. But it is a small town in its geography and mentality. The installation of traffic control poles at the intersection of Elizabeth and Eaton still rates a front-page story (with photo) in the *Key West Citizen*. Often called the Cape Cod of the south, in truth the island city is a good deal less sophisticated than its Massachusetts counterpart. I bathed off the pier at the Clarence Higgs Memorial Beach, along with a family of amiable pelicans. But the beaches of Key West are not world class. They aren't even Florida class. Scruffier than Cape Cod's Provincetown, with outcrops of pool halls and lowlife bars and great dollops of the lowest-grade kitsch, the old town is infused with a flavour of the 1950s, not least due to the overhead power cables and, when I was last there in 2002, the absence of Starbucks and internet cafés. Nor is the town exclusively bohemian. Middle America cruises to Key West for sunsets, Sloppy Joe's and Hemingway's house. Duval, the main drag that bisects the old town, is beyond redemption except at 6 or 7 a.m. when the trinket shops are closed and the last drunks are blinking in the tropical glare, and you can look down Margaret or Greve to a hopeful rectangle of masts. This place receives upwards of a million visitors a year. As for the cemetery a mile behind the front: one Key Wester, the people known as Conchs (pronounced *conks*) after the giant sea snails Calusa people ate, had commissioned for her late husband's tombstone the words, 'At least I know where he's sleeping tonight'.

Playboy once called the Green Parrot at Whitehead and Southard 'a true shitkicker bar', continuing to recommend it highly. The stygian Parrot now sells T-shirts bearing the legend, SEE THE LOWER KEYS ON YOUR HANDS AND KNEES. In 1931 Hemingway bought a house at 907 Whitehead with his second wife Pauline Pfeiffer's money. He wrote much of his best work on the first floor of the converted coach house in a cool, plain room in which a tiny Royal typewriter still stands ready for duty. The grounds are available for weddings, and during my visit a coiffed bride stood awkwardly on the lawn as tourists snaked past. You might ask why one would want to invest one's conjugal prospects with the spirit of the adulterous Papa. He spent more time drinking at Sloppy Joe's than he did in his house, and when the bar moved premises he installed its urinal in his garden. One day in 1936 a journalist on assignment for

Collier's approached Hemingway's stool at Sloppy Joe's. Her name was Martha Gellhorn. Oh, Martha. Key West is not Hemingway's town on my special map. He spread himself too thin to appear in any town, and he never wrote about places or invested them with meaning, not really: places didn't interest him. Tennessee Williams, on the other hand, kept a house on Duncan Street for more than four decades. He liked the sailors. A wooden pier was known, until its collapse, as Dick Dock. I've put it on the map.

Once, deep in the Colorado desert of southern California, my car temperature gauge read 106°F. The sun impaled both Coyote Mountain and the Borrego Badlands, where fan-shaped alluvial deposits called *bajadas* abutted folds of sedimentary rock. The light was as hard as the granite. Tindery blankets of faded purple spread over the foreground – phlox, I think. The horizon was not shimmering: it was too hot for that. Anza-Borrego is the largest state park in California, and to get there I had driven east from La Jolla, then northeast in a half-loop. Once the conurbations lay in the rear-view mirror, I passed at most one vehicle every half an hour. I said the park was deep in the desert, but in fact the centre is only fifty-five miles from the Pacific and eighty-seven from San Diego. When I got out of the car in search of a trail (there wasn't one), a lizard eyed me suspiciously from a baked rock. You had the sense that the whole scene might self-combust. The village of Borrego Springs, with 3,500 residents but no traffic lights, lies marooned in the middle of the park. ('Borrego' means bighorn sheep in Spanish.) As I walked the palmy streets, a suggestion of the *Mary Celeste* hung over proceedings. Everyone was indoors. Air stung the nostrils. On the edge of town, an RV park stood silent. I felt energy seeping out of my brain. Scientists agree that heat is not conducive to thought, because cooling the body requires more energy than warming it. Most of us have experienced the phenomenon.

Ever had a good idea on a beach? Cognitive function declines even with a moderate temperature rise. Unless, of course, one is acclimatised – especially genetically acclimatised. Cahuilla hunted and gathered across inland Californian deserts for many hundreds of years. In this particular region they ate the berries of the mesquite tree, the roots of which descend deep enough to find water even here.

Entering the Red Ocotillo café on Avenue Sureste was not so much like leaping into an icebox: it was more welcome than that. It seemed that if staff cranked the air con any higher, the entire twenty-seat café would turn into a freezer. The patronne, a cheerful woman in her forties wearing elaborate sand-coloured hair braids, approached with a menu and a pitcher of iced tea. Large windows bore faint smears of Windex. Chilled avocado soup came garnished with strips of tortilla and diced tomato. A blue-eyed man with a hat like a doughnut heard my accent as I chattered to the waitress. 'How are you coping with the heat?' he asked. We had a discussion about that. He was a roofer. It seemed hard to believe the job was humanly possible. Why didn't he fry? Among the studies I mentioned, one, conducted by researchers at the universities of Houston and Virginia, found that, over a year, the purchase of scratch-card lottery tickets that involved a series of decisions fell in line with each degree of temperature rise. As our bodies struggle to maintain a healthy internal heat balance, they consume resources otherwise available for decision-making.

The sand-haired one approached. 'Would you like a coffee?' she asked brightly. I couldn't make up my mind.

Then there are the human extremes, the figures cast against these freakish landscapes. In Alaska I had met loners and eccentrics in flight from the Lower 48. But in Lancaster County, Pennsylvania, in 1995 I spent a month with an Amish community. I was thinking of writing a book about the Amish, as their commitment to a spiritual realm and its elevation beyond any material equivalent held deep fascination. I liked in particular the Amish actual groundedness in soil, a characteristic generally perceived as the antithesis of the transcendent. In the end, I decided I could not square the circle of trust and confidentiality. Of course, I could change names and appearances. But I could not obfuscate without obscuring their truth, which is what I was trying to reveal. Or so it seemed at the time. In the aughts, when I was writing about Denys Finch Hatton, I came across another American human extreme. I was researching Finch Hatton's lover Karen Blixen. Peter Beard knew her. I approached him, and he invited me out to his house overlooking the ocean in Montauk, on the tip of the South Shore of Long Island in New York State. Beard, then in his late sixties, was a celebrated photographer who moved in glamorous circles. He was once married to the first US supermodel. He met me from the Hampton Jitney I had ridden from Manhattan, weaving across the road gripping a plastic beaker of vodka Clamato. Beard made collages from prints of his East African images which sold for many hundreds of thousands of dollars at Christie's in New York, London and Berlin. At the Montauk studio, a team of amiable assistants were on their hands and knees pasting photographic prints, some of Blixen herself. I talked to Beard, who had lived in Kenya, as we ate lunch with the team at an inn. At night Peter and I stayed alone out at his house on the Point (his third wife, Nejma, was in Manhattan). There was no food of any kind in the kitchen bar half a packet of stale raisins which I found and ate in the middle of the night while Beard channel-hopped in front of a television screen of galactic proportions and whispered things like 'the kiss I gave you' into his phone. Before he rose I walked in the woods on the headland, and in April 2020 it was there that they found his body. He had been missing for three weeks. Later that spring a publisher commissioned the journalist Graham Boynton to write a biography of Beard. Boynton interviewed me recently and revealed that I was the only woman he had come across at whom Beard did not make a pass. At last I had something to record in *Who's Who*.[1]

Cheap energy has been the single most important factor in the environmental landscape of my travelling lifetime, and the American energy frontier in Alaska represents the epicentre of tension between preservation and development.[2] It was impossible to ignore either as a writer or as a member of the human race, so I went. I've said I like the extremes of the USA. Is anywhere more extreme than Alaska? It was the ultimate manifestation of the America Whitman celebrates, 'a teeming nation of nations'. I recognise that heterogeneous land, so alien to manufactured MAGA unity. By far the largest state, Alaska is six times the size of Great Britain. It has a hundred thousand glaciers and, with only a twentieth of 1 per cent of its land developed, only a little over one person per square mile. State licence plates bear the slogan THE LAST FRONTIER. But even by the standards of a region that considers itself isolated, Arctic Alaska is an isolated wilderness – one of the last true ones on the planet. And what of its oil? Crude first flowed through a pipe on 20 June 1977. Twelve years later, on 24 March 1989, the 987-foot supertanker *Exxon Valdez* ran aground on Bligh Reef with 1,263,000 barrels on board. Remember the television pictures of oil-stricken sea otters and tar-strangulated gulls? The Exxon PR machine was ready. The company released footage of 250 otters being flown to rehabilitation centres where they dined on crab; to the joy of the world, all 250 survived. It cost Exxon $90,000 per animal. The rest of the oiled otters died.

Hearts and minds are one battle. No side will ever win the war. The Big Melt will win in the long run, when we are all dead. (To his credit, on his first day in office in 2021, Biden suspended drilling in the Alaskan Arctic.)

---

1. I have also done aerobics classes on seven continents.
2. Writing in 2022, I have outlived even this defining feature. Rising energy costs and fears over national fuel security dictate the headlines now.

Because of oil, one of the greatest roads on the planet cuts through Alaska – the Dalton Highway, the only land route to the Arctic Ocean, the all-time über-road that met every engineering challenge. In 2007 I took a trip along the Dalton and followed the Trans-Alaska Pipeline to the Arctic Ocean. Any pipeline had to cross land occupied, but not owned, by indigenous peoples, and in 1971 the Alaska Native Claims Settlement Act passed into law. Much has been written about the way in which the Act transformed native Alaskans into capitalists. Slope Borough, an Inupiat initiative headquartered in Barrow, taxed the Prudhoe Bay oilfields, and Barrow became the richest settlement in the US, eclipsing Key West in its pomp. America had hit on a capitalist solution to the problem of indigenous exploitation. But bombs could not have destroyed a culture as effectively as the Act. John McPhee, one of the best writers on Arctic Alaska in the seventies, noted that all Hungwitchin people had to do to get their share of the money was 'turn white'.

The Arctic Science division of the University of Alaska Fairbanks had offered me passage on a delivery run from their North Star Borough city, where the Dalton begins, up to the Arctic Ocean. Behind the wheel of a truck marked with the university logo sat Jeannie. She was an Idaho-born outdoorswoman, once an oil worker on the rigs, now an employee of the university maintenance crew. Travelling in a pickup of a size never sighted in Europe, Jeannie was heading to Deadhorse, the service town adjacent to the Prudhoe Bay wells, where she was due to collect a consignment of hauling equipment. Meaty as a wrestler, with cropped hair the colour of hay, she was an old hand on the Haul Road, as truckers call the Dalton, and a steadying presence when eighteen-wheelers hurtled past catapulting rocks at our windscreen.

We had pulled out of Fairbanks shortly after dawn, bouncing through boreal birch forest, stands of aspen, evergreen spruce and alder. Tassels of sedge flecked valley floors golden with cottonwood, and on the low ridges, against a background of glowing green, the rich cream of reindeer moss reflected the sun's rays. Initially nervy and reticent, Jeannie thawed as the population dropped away, and besides extolling the wonders of the Alaskan wilderness, she sometimes turned towards me just to flash a smile. Sitka spruce, elsewhere so mighty, shrank as the latitude – monitored on the GPS on the dash – crept towards sixty-six. 'Howdy, buddy,' a male voice crackled over the CB radio. 'I'm right behind you, and I'm not gonna be able to stop as I slide down the hill over the ridge ahead.' I looked up from the passenger seat to see the rear-view mirror filled to its frame with a monster fender powering towards us in a nimbus of coffee-coloured dust. 'I'd be grateful,' drawled the voice, 'if you could pull over to let me pass.' That would be a yes, then. The Dalton Highway was a macho stereotype: the biggest got priority, just for being big. If you disagreed, you got flattened. A glint of the Yukon flashed through birch trunks. The two-thousand-mile river was once the focus of life in the Alaskan interior, and gave access to the Shangri-La of the Klondike. To indigenous peoples, it was a lifeline to fresh hunting grounds when the weather changed or the caribou altered their route. In the seventies the plan to ford the Yukon in Alaska presented civil engineers with a challenge. Air temperatures ranging between +32°C and –50°C required a differential of two feet to allow a bridge to contract and expand. The steel box-girder construction that now links both sides of the Dalton has the pipeline strapped to its side. As you approach, you can see the slope. Since 9/11, the combined presence of bridge, pipeline and riparian Pump Station Number Six have necessitated a heavy security programme – imagine the value, to a terrorist, of cutting off both the oil supply and vehicular access to Prudhoe Bay. Security personnel maintain a twenty-four-hour vigil close by the north shore stanchion; when we passed, a guard on the day shift was eating a sub in the cab of his van.

The temperature dropped, an Arctic zing supplanting warm

currents. Something in America unclenched. We stopped to wait for a young grizzly to get up from the middle of the road. He was licking his front paws. The pipe had been part of Jeannie's life for a long time. 'See the zigzag?' She pointed at the silver zeds on the hillside ahead. 'Protects against earthquake damage. Had a 7.9 Richter scale 'quake here in '02, and neither the pipe nor the big Yukon Bridge so much as bent an inch. I was here!' Alaskan permafrost might be in excess of two million years old. (Unlike most of the Canadian High Arctic, during the last glaciation Alaska had no ice sheet lid to keep out cold air.) What have those cubic miles of frozen earth locked in? For pipeline engineers, the solution to permafrost was to elevate the pipe on stilts for 420 out of the total eight hundred miles. The stilts were fine – noble even – but the hillsides covering the buried portions resembled lurid green ski slopes. The pipe itself had a certain sleek, technical beauty. Overall, it wasn't ugly. It was unobtrusive. Two miles before the Arctic Circle ice skimmed the hillsides and beaver lodges damned the creeks. At the Circle itself, marked with a sign, we stopped to eat the Alaskan picnic Jeannie had prepared for my benefit: sockeye salmon jerky smoked in her back yard (she called it 'squaw candy'), home-baked bread, wild cranberry muffins and a flask of dark coffee. Jeannie could kill, skin, gut and quarter a caribou without assistance. Solitude and the back-country had proved more reliable than human company, and she played a man's game on the Haul Road. She had even made herself look like a man.

Coldfoot, a destination touted as the major settlement on the road, was named after gold prospectors who got cold feet. During the mini-boom of 1902, Coldfoot Camp had seven saloons with wind-up phonographs, a post office, a jail and ten whores, the latter referred to as sporting girls. We drove into a car park surrounded by six low buildings made from abandoned construction containers. This turned out to be Coldfoot. Inside the surviving saloon a plaque recorded a temperature of 97°F (36°C) in 1988, and below it a sign noted that the following year Coldfoot had experienced −50°C seventeen days in a row. Jeannie and I both ordered the Special, pork belly soup, which was excellent. In the afternoon the valleys deepened and the forest yielded to semi-iced North Slope tundra. On the steepest hillside harriers and rough-legged hawks circled above a flock of wild Dall sheep. Jeannie gunned the truck into the foothills of the Brooks Range, the seven-hundred-mile limestone uplift which arcs across Alaska, the end of a spine that starts with the Andes and surges north as the Mexican cordilleras and the Rockies. Here its mountains mark the border between the Arctic North Slope and the Yukon River Drainage – what some call the Arctic Divide. Behind the first row of low hills, blunt eight-thousand-footers rose in waves, deeply incised by broad, U-shaped glacial valleys.

Jeannie had become more voluble as we approached journey's end. She had started to shout the name on each sign. The road builders were keen on signage, and the forks of the rivers were capable of infinite subdivision. 'The West Fork of the North Fork of the Chandalar River!' Jeannie yelled. As an act of preparation for the descent waiting over the highest pass, she got out to lock the hubcaps. On either side, the highway dropped away. Jeannie climbed back into the cab. The sun was setting, painting the sky rosy pink. When the valley was in shadow, level sunbeams continued to pour onto the white higher slopes in an effect of ravishing beauty, and the ice reflected and refracted purple light like cut crystal. Then all except the tips turned white, and we experienced the transfiguration of alpenglow. When the shadows crept higher and submerged both slopes and ridges, alpenglow lingered on the highest peaks, until eventually these too were quenched, points dying like stars. We had crossed the Arctic Divide.

'What keeps you coming up here?' I asked Jeannie after a period of silence in which we both contemplated this ethereal wilderness. 'It's not what's here,' she said. 'It's what's not here.'

Nothing in the Arctic defiled Whitman's beautiful whole, 'For every atom belonging to me as good belongs to you.'

Canada has 151,000 miles of Arctic coastline, much of it increasingly ice-free, compared with a US tally of a stingy 1,060 miles. When a countrywide survey quizzed the public on what it means to be Canadian, the majority of respondents cited 'not being American' as the primary characteristic of nationhood. But as a secondary badge of identity, Canadians pointed to the existence of 'our north'. 'We are a northern country,' its citizens repeatedly asserted in the poll. 'We have our Arctic.' Only a hundred thousand out of 33 million Canadians reside north of sixty. Many countries are burdened with too much history, but as Prime Minister Mackenzie King said, Canada has 'too much geography'.

I transited once in Iqaluit, a full 1,300 miles from Ottawa, on my way to a geoscientific mapping camp on Southampton Island where I was to observe the àdvance guard of mineral extraction on the job. The sky lay on Frobisher Bay tight as Tupperware, and chained dogs howled all night, lupine whines that carried far over the water. The temperature hovered at five degrees above, and in the mornings the sky was striated, like bacon. Two wide paved roads forked through sections of Iqaluit's small, semi-urban sprawl while sand roads gave up in roadblocks of mud and ice. The jostling pack of the bay dominated the scene – that, and the beach in the heart of town, damp sand crowded with rowing boats, motorised dinghies, locked wooden storage huts, ambulant dogs and leaking mounds of rubbish, signature of the inhabited Arctic. Giant signs appeared in three languages to reinforce the Canadian-ness of the nation's newest territory: French and English, and Inuktitut in both the Roman alphabet and in syllabics, a script invented by missionaries. (The proselytisers gamely tackled metaphors: the first Eskimo Lord's Prayer asked Him to give us this day our daily seal.) Stooped old Inuit sat around, children skidded on home-made skateboards and teenage girls plugged into iPods drifted around in *amautiit* duffels, a baby peeping from every hood. Inuit had exploited their environment in collusion with the Arctic for many centuries, curing seal hides with spittle; catching bowhead with bone harpoons; conjuring gods and spirits from the ice to invest the universe with meaning. The battle for their diet was among many still being waged since the white man dragged his 'northerners' into the cash economy. It had cycled dismally from orders not to eat the loveable whales to orders reversing that order. That proceeded to advice not to eat whale meat after all as the white man's sturdy flame-retardants had stormed up the food chain to contaminate the flesh of Jonah himself. Later in the summer, at a supermarket in Coral Harbour, I saw where the field of battle now lay. The spacious, windowless aisles pumping out music composed to engender a mood of high-spending relaxation could have been anywhere in the developed world. But fresh food was limited to Californian peaches so withered it seemed inconceivable anyone would ever buy them, coal-black bananas in a similar condition, shrivelled grapes, mouldy cheese and cartons of milk well past their sell-by date. Of processed food, there was an abundance. Gloopy non-dairy cheese dips in plastic tubs, packets of flavoured sugar crystals for the preparation of fizzy drinks, Pot Noodles, instant mashed potato, snack-size 'pepperoni' sausages, lurid Miracle Whips, fluorescent candies, and more, much more. It seemed that every processed item in the world had struggled up to 63 degrees north, up across the latitude lines and the treeline, over the tundra and the pack ice, over the wastes of Hudson Bay, through the indigenous settlements battling for dignity, down into the overheated, lino-floored, echoing corridors of a giant supermarket provisioning eight hundred residents of Coral Harbour. And as if that weren't enough to make them obese, close to the checkouts a worker was unpacking a consignment of LCD screens the size of caribou and marking them with prices that were double or triple those below sixty.

People often ask me if I am afraid and I generally reply that the open road holds no terrors and that I experience fear, as I have said, only among John Lewis drapes. But this is not quite accurate. I was afraid once in the Canadian Arctic. I was a guest at the government-funded camp I mentioned on Southampton Island, a rather bleak interlude when it rained half the time and blew a gale the other half. The team were investigating what the land could produce in the way of natural resources that might stimulate an economy. The hydrocarbon and mineral potential of the Hudson Bay region has become an investigative industry of its own. The joke used to go that a typical Canadian–Inuit family consisted of a father, a mother, two children and an anthropologist. Now a geologist has replaced the anthropologist.

At about ten o'clock one morning the camp chopper had left two scientists, a field assistant and me a couple of miles to the southwest of camp, where we started out on foot across a plain surrounded by low hills and glacial outwash deposits. The sky was flawless that day for once, and mist rose from a livid blue lake. We had walked a mile from our first station and just arrived at our second. It was warm by northern standards, and our backs had been sweating under packs. As we began getting out observational instrumentation, Joyia, the lead geologist, stiffened. She said, 'Bear.' I pulled at my sock, which was rubbing under the lip of my boot. Eight hundred yards away, a polar bear was loping up an escarpment. We had been downwind of him for an hour, so he must have smelled us, and after a few minutes he lifted his snout in the air and began to circle. Eight hundred yards might seem like a long way to a reader sitting in an armchair. When there is nothing between you and a bear that can outrun you, eight hundred yards is shortish. Joyia cocked the gun, and the three of us loaded anti-bear firecrackers that allegedly frighten ursids into running away (an improbable outcome, I always thought). Half the world's polar bears hunt in Nunavut. But only one mattered. We called camp on a sat phone, asking for a pilot to come and get us. (Weeds!) The bear meanwhile completed a quarter circle before disappearing over a ridge. We kept vigil, and chatted. Joe, the other geologist, looked at rocks; it was difficult to say whether his nonchalance was studied. Joyia alternately scanned the horizon with field glasses and fiddled with her Brunton compass. I tried to think of something to do, my mind fixated on images of motherless sons. Then we heard thwok-thwok-thwok, the holy sound of a helicopter.

I spent a long time immersed in the geopolitics of the Arctic when I was wrestling a book to the ground. Many years on, we know that the loss of its ocean's ice cover might prove to be the most important statistic, not the extent of its hydrocarbons. I wonder if ultimately, when the last glacier has melted and nationality is a trivial thing, the values of the peoples who once cured seal hides with spittle won't have won after all.

6

Its Own Map

Latin America

If I could find one person, I mean out of everyone I've ever met, it would be José Gomez. He was my boyfriend in Chile. When I met him in the Elqui Valley my legs were sticking out from under a jeep, as I was changing a tyre. He stopped to ask if I needed help, and one thing led to another. It was a romantic place to meet. A kind of supernatural allure underpinned by atmospheric physics shimmers over the Elqui Valley. Because of cloud-blocking Andean slopes to the east and air-scouring desert to the north, the night sky is almost uniquely clear, and since the sixties the most technically advanced observatories on the planet have built scopes in the valley. I went to visit one, expecting to press my eye to the end of an enormous black tube. In fact the telescope resembled my house sheathed in hard white plastic, a Christo installation relocated to the Atacama. The South Pole and Hawaii also benefit from that exceptionally thin atmosphere required for stargazing, and later, at the Amundsen–Scott research station at ninety south, I saw AST/RO – the Antarctic Submillimeter Telescope and Remote Observatory, at that time the most penetrating scope in the world. It weighed six tons, had a 1.7-metre diameter off-axis and detected short wavelengths known as submillimetre radiation. The lead astronomer told me those details as we inspected the machine on the roof of a custom-made building on steel legs. 'It can look into distant galaxies,' he said as his beard iced up. Apparently there's an empanada place called Halley's in the Elqui Valley now.

Numbers three to six of my Chile notebooks, not-quite-disintegrating hardbacks of a size that can fit in a generous pocket,

reveal José's handwriting spelling out a word and its etymology. 'Chirimoya. *La fruta que te gusta.* Quechua word. Only grow at altitude. Don't crush the seeds!' He had appended a cross-sectional drawing. José was an environmentalist at a time when only a few ploughed that field. He had a degree in something to do with it. He also acknowledged the spiritual realm and its importance in the temporal one. We often had conversations about this, though, looking back, I wonder how often my poor Spanish meant I got the wrong end of the stick. He was keen on the tantric, though that's quite enough about that, and had some notion of causal relationships between events that superseded anything rational. My diary notes one exchange which ends rather unsatisfactorily with me asking, 'So you mean it's like magic?'

The notebooks themselves (narrow feint-lined, no margin) are an undistinguished, classless brand, not the Moleskine Chatwin fetishised. I used to buy them – they have a black cover and red spine – at my local stationer, since evolved into the seventh coffee shop in the small neighbourhood (the antonym 'devolved' would be more appropriate). I would leave the purchase till the week before a big trip, as it constituted a longed-for pleasure compared with tasks of the vaccination and flat-lease variety. On one occasion, the stationer said he was out of that type of notebook as the supplier had stopped manufacturing them. I stood looking out at the traffic of South End Road. Was the trip off, then? How could I manage with another brand at this late stage? At any rate I did manage, and the notebooks stand platoon-like in a row on an office shelf now, each marked like a soldier , as I mentioned.

José and I travelled together mostly in the north where we had met, but on the coast rather than inland. He had an old canvas tent with bits of marijuana stuck to the groundsheet. There weren't any campsites, except one in the Parque Nacional Bosque Fray Jorge in the Talinay coastal cordillera. Fray Jorge was famous for its fog, the *camanchaca*, formed from marine stratocumulus. We watched it roll into the forest where soapberries and Peruvian peppers and olivillos sucked up its moisture. I first saw a penguin in its natural habitat with José not far from there. We had asked a fisherman to drop us and the tent on one of the uninhabited Choros islands and hiked to a colony of Humboldts. Later, in Santiago, I had arranged a trip to La Moneda, the presidential palace where US-financed Hawker Hunters murdered the democratically elected Salvador Allende in 1973. The palace (the name means mint, and the *moneda* had produced coins until 1929) had been closed to the public for years, and José remarked on the fact that the authorities ushered in *cualquier extranjera*, 'any old foreigner', while keeping Chileans out. I couldn't let this go.

'I'll ask if I can bring you. You could be my photographer! Do you have a camera?'

'Yes! My sister won one at a school fair. I could borrow that.'

This apparatus seemed unlikely to convince even the most nonchalant apparatchik. I had another idea.

'You could be my translator.'

He couldn't speak English, to speak of. I recall with fondness the Laurel and Hardy routine in the palace corridors as a crusty administrator reeled off boring historical detail not involving 1973, which I translated out loud for myself to vigorous nods from José.

We wrote once or twice after I went home, but there was no pretence on either side that return trips were on the cards. It wasn't practically possible. Of course love will find a way, but, well, you know.[1]

Every day in Chile, a country twenty-eight times longer than it is wide, I wondered why they didn't manufacture maps to spool

---

1. A girlfriend of mine in her seventies spent lockdown hunting three swains of her youth, two of them in foreign lands. The pursuit, as so often, was the best part. The former lovers, whom she eventually met again on Zoom, were no longer handsome young men who prised open shellfish straight from the rocks like my José. 'They were all jowly and bald and fed up with everything,' my girlfriend reported.

out on rolls like lavatory paper, with a spigot in the middle. I once interviewed the national television weatherman in Santiago. This heroic figure was obliged to say, every day, 'Tomorrow it will be hot at the top, cold at the bottom and warm in the middle.' They had divided the country into twelve regions named not after glaciers or cactus-studded deserts but numbered in Roman numerals like chapters in a textbook. It was hard to summon in words an endangered *jubaea chilensis* sprouting from the pointy triangular hills of northern Maule when you had to add *Region VII*. The president's office announced the creation of three more regions in later years, and in 2018 dropped the Roman numerals. Once again, history has swallowed the record of my modest travels. My interest in maps is itself obsolete. The past is another cartographic country; of course actual maps aren't superior to their digital descendants; and anyway, to paraphrase Les Dawson, the way things are going, the good old days were last week. But still, maps are woven into the fabric of my memories. (I've mentioned the Limit of Compilation.) Among my papers there is a map of Chile's Isla Mocha annotated by a government census-taker. I hiked round the island with him on Census Day in 1992, asking Mochanos how many children and microwaves they had. Most of the island did not yet have mains electricity. We don't need paper maps now, but indigenous peoples have long been more evolved. Inuit hunters carried extraordinarily detailed maps in their heads, and I have on my desk a copy of a map Iowa chief Notchininga drew in 1837. It depicts, in sepia lines of what turns out again to be unfathomable accuracy, the drainage system of the upper Mississippi and Missouri rivers. Notchininga presented the document to the federal government in faraway Washington in a desperate plea for restitution after settlers stole Iowa land. Yet on a map of the frontier published in Virginia a few years later, on the left tracts of apparently empty land appear horizontally banded in pink, green and yellow, beckoning young men west. The stripes on the map weren't really empty, but to people seeking a new world, they represented possibility. During Covid, I thought a lot about the *possibility* of unmapped territory. How were we to navigate these strange empty spaces as our own frontiers drew inwards rather than moved outwards? Like Coleridge's albatross, a map – Google Earth or one that folds up, or, more usually, fights back – reassures us that there really is a way out. It gives hope, when we know that inside we are navigating by the stars. A journey is the oldest symbol of the lone human endeavour to battle through, and a map gives us a purchase on it. King Nauplius, holding his torches aloft on an Evian cape to lure enemies to their doom, highlights the seduction of wrecking rocks on all our journeys.

I learned as I travelled down Chile that despite its long, thin shape, the country is more of an island than any island I knew, including Evia and Great Britain. Geography defines it: the driest desert in the world, the Andes, land that collapses into glaciers, and the Pacific. And geography is a driver of history, as you can see in the examples just cited of Evia and Great Britain. When I was writing my second book, about a six-month journey from the top to the bottom of Chile, I had imagined the latter to be Cape Horn. To get to the Horn I had inveigled myself onto a supply boat servicing a cruise ship. The small firm operated out of Puerto Williams in Tierra del Fuego and its obliging owner was happy to take me along, as there was plenty of room on the boat and a crew of only ten. (I had met him when I nipped out of my guesthouse to post a letter.) We chugged through channels where Yaghan people once dived for shellfish and past thin lines of beech which southerlies had bent into alphabetic configurations. Inevitably, a gale blew at the Horn. It was dark, and only when the crew mustered did I realise that the long, narrow box covered in a candlewick bedspread on which we had been playing carioca was a coffin. All monster cruise ships (they were not as big in 1993 as they are now, but this one, crouched in the black water, looked titanic) carry coffins. The vessel had run out. Two days later I flew back to Punta Arenas from Puerto Williams on a sixteen-seater commercial plane. Wedged in the aisle, a familiar shape. Next to it, a weeping woman in an unfunereal orange safari suit. The corpse and the widow were on the long journey, his last, back to Milwaukee. My friend at Cunard told me it was unusual. 'Most often the widow carries on,' he told me. 'She says, "It's what he would have wanted."'

After all that, I had still not done enough to complete my portrait of contemporary Chile. I had noted, making my way south (the working title of the book was 'Keep the mountains on the left') that at the foot of every map, even those sewn on to the uniformed arm of a member of the Agrupación Nacional de Boy Scouts, a triangle hung, like a slice of cake. This was *Antártica Chilena*. The country was one of seven land claimants. No other nation recognises land claims. It was not yet anxiety over hydrocarbons locked (for now) beneath a mile of ice that had fostered claims. It was amorphous geopolitical ambition, or what we now know as FOMO – fear of missing out. This made sense to my nascent understanding of the thin country. Chile declared itself independent of Spain in 1818, which was yesterday, in Les Dawson's view; I began to understand its near-perpetual state of aggression towards its three neighbours over fragments of disputed land. Chileans had found their new world and were fighting to keep it. The issue of territorial sovereign unity was so important that the government had passed legislation banning the publication of any map which did not depict the slice of cake. So I had to go to the Antarctic in order to complete my portrait of contemporary Chile. Was there to be no end to it?

People often ask me how, on the road, I make decisions about where to go. You just know, and you instinctively pick which mountains to climb, usually having to fight internally with yourself to acknowledge it, as the ones you have to climb are the steepest. Often it turns out you don't 'just know' after all as I have said, and that you have wasted time, or worse. All I can remember is that I had to get to the Antarctic, and that I regularly walked for two hours among the volcanoes of Region VIII to a phone in an exiguous shop, a device which usually wasn't working (one had the number Chacabuco 1, but only accepted incoming calls), then walked a further two hours to another phone in another shop in order to maintain a series of conversations with a Chilean general who had the power to inscribe me on the manifesto of a military supply aircraft from Punta Arenas to the Chilean Antarctic station. That went on for many months, the conversation set to continue until one of us died or the Atacama cactus that blooms once a decade had flowered twice. When I finally jumped from the metal folding steps of a CASA CN-235 and looked over an ice desert bigger than the continental United States, land that remained gloriously unowned, whatever laws senators cared to pass, my next book appeared in my head, like José's magic.

Immersion in Chile had prejudiced me against los Arrrrgentinos (Chileans mock their rolled rs). Neighbours the world over are enemies in jokes – Romanians tell them about Hungarians, Lithuanians about Estonians, Britons about Irish. But Chileans resent Argentina for more than I could begin to know, and as the Falklands conflict was fresh in the collective mind they liked the British – their enemy's enemy – a situation on which I capitalised. (Generally it remains advisable everywhere in South America, not just Chile, to advertise one's Britishness as soon as possible, in order to disassociate from the Estados Unidos.) A shortcut across the border in the south, where traversable land disintegrates on the Chilean side, would be disloyal. So I doggedly threaded my way down on cargo vessels, disgorging at Puerto Natales in Última Esperanza. I did not set foot in Arrrrgentina on that trip, or for twenty-five years thereafter. Then, when an assignment presented itself, I went.

I remember the ghost of Borges. At a fuggy café in Recoleta, over tiny cups of cortados and flaky facturas filled with quince

jelly, I glimpsed him looking through the steamy window, close
to the glass, as he was almost blind. Borges was obsessed with
maps, and in 'Del Rigor en la Ciencia' imagines a life-scale map of
an empire, which is of necessity the same size as the empire. (He
got the idea from Lewis Carroll's 'mile-to-the-mile' map of the
world, never yet spread out as farmers objected on the grounds
that it would block the sunlight, so the men who had made it 'use
the country itself, as its own map, which does nearly as well'.)
Many Borgesian stories are set in the cafés of Recoleta, the *barrio*
in the north of Buenos Aires where the writer and his mother
lived on Quintana Avenue, round the corner from the National
Library on Mexico Street. Borges served as director of the library
for eighteen years and the institution too appears often in the
stories, a brooding character in itself. Jorge Francisco Isidoro
Luis Borges Acevedo (1899–1986) was blind by his mid-fifties,
like his father, grandfather and great-grandfather before him.
They also shared a womanising trait, and Borges once said that
his father had tried to pick up his own wife, Borges' mother, in a
Recoleta street after they had been married for twenty years. (He
didn't recognise her.) When Argentinians re-elected Juan Perón
in 1973, Borges resigned his library post. 'If their posters and slo-
gans again defile the city,' he told an interviewer in Buenos Aires,
speaking of jubilant Perónistas, 'I'll be glad I've lost my sight.'

Borges conjures grillwork above pocked stone arches and the
Gas Lamp Coffee House where he sipped coffee and absinthe
while three horses and a delivery wagon waited on Alsina Street.
The narrator of 'The Waiting' has a tooth pulled at a dentist in
Buenos Aires' Once neighbourhood and the nearby hotel displays
peacocks on its crimson wallpaper. Borges allows the reader to
see these things, even when he can no longer see them himself.
He was born in a city of sixty-eight thousand. The population of
Buenos Aires is now fifteen million. But what Borges describes
has endured. In the Cementerio de la Recoleta fresh roses and
branches of flowering ceibo lay on the tomb of Eva Perón in the

Duarte family mausoleum. The midday sun glinted on dark metal plaques honouring Evita, and a trio of pilgrims paid homage, setting on the mossy step a green glass bottle holding a wilted *bandera*, the blue-and-white Argentinian flag. I glimpsed Borges again there, shoulders hunched, hurrying away to his streets, those radiating from swanky (then and now) storefronts, past 'spotted plane trees, the square plot of earth at the foot of each, the respectable houses with their little balconies, the pharmacy alongside, the dull lozenges of the paint and hardware store'. In the same pharmacy windows, porcelain letters spelled out the name of the patron, just as they did when Borges peered through the plate glass. His stories express nostalgia for a Buenos Aires you might think had vanished with the horse-drawn delivery wagon; the writer himself, seventy years ago, felt 'certain national values' slipping from the old criollo *barrios* in the post-war double boom of economic development and immigration. But the heart of a true city never stops beating, on the page or in the hard red earth.

A sheaf of children's drawings of Chile travelled home with me, and, seeing them all together pinned to my office wall, I note that as the *regions* tumble the Andes appear as a line of triangles down one side: sentinels, guardians, protectors. Argentina shares most of those mountains, but it is too big, and like Canada has too much geography, for the peaks to play a central role in the national myth. But they are there. I went to find them in Salta Province, 1,460 kilometres northwest of BA. The mineral silence of its puna heartland moved me. Puna, a Quechua word meaning 'a cold and remote place difficult to live in', is formed by an inter-connected system of volcanic cones, giant lava outcrops and the elevation of the Andes, all sliced through like hard-boiled eggs by *quebradas* (a particular type of Andean gorge) and *vegas* wetlands (as in Las Vegas, unlikely though it seems), the latter appearing in startling bursts of teal green stripes between bleached hills.

Over three days, I passed nine vehicles. Three hours from Salta City, at Susques, the last substantial village before the Chilean border, I turned onto Ruta 40, a quiet five-thousand-kilometre epic that shadows the mountains from the top of Argentina to the bottom. The volcanoes north of Susques are the highest peaks in the world beyond the Himalaya. Twisters – a kind of desert tornado – spun on the dry white horizon between them. Salt flats and dry salt lakes (*salinas* and *salars*) pop up all over the puna. Polygonal rills of salt-like sand characterise the flats, and when you kick a piece of crust it tinkles like glass in the sharp silence. Some salt lakes extend for many kilometres, six-hundred-metre-deep sodium chloride blotched by the occasional pool of Hockney-blue water. Aqueous mirages wobbled (an optical effect of salt reflecting sky) and dust fizzled through dry air, scoured from ancient clay formations sparkling with chalk crystals. At night, as the temperature crashed, the sky faded upwards into weak violet heights.

The figure of Pachamama, a pagan mother-earth deity grafted on to Christianity and dear to Andinos, still flourishes up in the northwest. At a puna shrine, Catholic baroque met primitivism next to a ziggurat of empty beer bottles. Nuns had removed the statue of Pachamama inside the shrine, replacing it with a big-eyed saint. I was there as the annunciatory glow of the rising sun appeared in the east, dissolving shadows on the Andes. I remember – and this doesn't appear in my notebook – that there was something incontinently human about the clumsily positioned bricks of the shrine. I must have wanted an otherworldly scene when I wrote my notes, but as I see things now, it is the human hand that brings meaning. In the only shop in Tolar Grande, I tried to buy for my sons one of two footballs hanging from the ceiling. The shopkeeper busied himself with a rusty stepladder, and eventually said he could not sell me the item as the balls had been there so long he couldn't remember the price. The black-and-red notebooks, I was beginning to learn as I excavated my travelling past, were reliable witnesses to the woman I was then. But a cross-examining barrister might ask, 'Do you still trust her?'

There is melancholy in everything I wrote about Argentina. Was it because I was so much older than I had been west of the Andes? I turned thirty in Chile, still imbued with the sense of invincibility associated with the young (the lucky young, not those living permanently in the shadow of actual death). I was fifty when I first went to Argentina, the age a writer referred to as 'the retreat from Moscow'. I had learned to live with a constant hum of anxiety, as most do; some say it's a result of having children, but I don't think so. Yes, as a young woman I was angry all the time in Chile at the tragedy Pinochet and his supporters had imposed on a country I came to love deeply. In 1992 memories were fresh. I met people whose fathers had been pushed out of helicopters over the peaceful Pacific. One girl was ten when she heard the night-time knock at the door. 'Mi mamá,' she told me, 'se enojo porque no se puso los calcetines.' They had taken her father away (for ever, as it turned out) and her mother was annoyed because he hadn't put his socks on. Mine was the righteous anger of youth.[1] In Argentina there was instead, as there must be in one's fifties, a bone-weary revulsion: at corpses piling up in foreign lands, at juvenile ear conditions in London's Tower Hamlets thought to have been eradicated in the Victorian era, at my own wretched stupidities, and yours. Beckett described sorrow becoming, 'something you can keep adding to all your life . . . like a stamp or an egg collection'. In the old days I would have added, 'or stamps in a passport', but we don't have those now.

So what of *return*? People often ask too if I ever go back. On my first return visit to Santiago I stood at a sixth-floor hotel-room

---

1. Elsewhere, in the months I was in Chile Black Wednesday marked the suspension of the UK's membership of the European Exchange Rate Mechanism and the Conservatives won a fourth term. Americans elected Bill Clinton their forty-second president. A cyclone killed more than 135,000 Bangladeshis. The World Wide Web came into existence.

window, moved once more, watching Chile's first female president below waving at crowds in the street. The world changes. But does one stand still as it spools by? The observer surely has changed too, despite Ian McEwan's comment that nobody changes much beyond the age of two. *Return* is about toggling both – what you see altered, and what you have become. I went back to Evia in 2018 for the first time in more than three decades and failed to reconnect emotionally or even aesthetically. The magic had gone. But looking back, I was going through a bleak phase of my life – a kind of puna – and it was impossible to toggle my inner life with the marble threshing floors of Kavodoro. If King Nauplius was there diverting the triremes, he wasn't holding up a torch for me. I was on the rocks myself.

Some things vanish with the impetuosity of youth, irrespective of experiences. I said I would give anything to sit down with José again. I'd give anything more generally to get back those unanticipated bursts of emotional intensity during my younger years on the road, the splashes of passion in Dufy-bright colours on the landscape and in the notebooks. I was amazed, re-reading my ugly handwriting in the black-and-red volumes, many of the pages literally splashed with unidentified beverages that had fuelled what passed for industry, at how many names I failed to recognise. 'Went to Nikos' sister's wedding in Spalia, where they served batida de coco.' 'Felt sad when Felipe left.' 'FUCKING CARLOS SPOILED THE WHOLE TRIP.' I bet he didn't.

I experienced a kind of redemption after that sad visit to Evia when I went back to Chile for a third time, on assignment in the far south for *Vanity Fair*. The interior puna had fallen away, and I could see clearly again, or as clearly as one ever sees anything, as Borges recognised. I had learned by this time that you really do take yourself with you when you run across the sea, and that there is little point in fostering illusions. Chilean Patagonia was still an awfully big adventure though. In my photographs of Chacabuco Valley, a flock of upland geese rise in unison and

hover over the steppe like washing flapping on a line, and the
Andes behind ripple in morning mist. It was the end of the world,
almost: the place where mountains sink into water. I remembered
that zone where the land splinters, obliging ferries to take over
from the road. This was lonelier than the actual puna by far, and
Alaskan in population density.[1] The name comes from *patagon*,
which means 'big feet' in the language of the sixteenth-century
sailors who first encountered Tehuelche. The mariners marvelled
at how tall the hunters were, which is mystifying, as other indigenous groups were short. (And still are. In January 2022 I was in
Chiapas, the state with the second highest indigenous population
in Mexico, and everyone was tiny.) The canoeing Yaghan were
exceptionally small. They were the people who lived off shellfish
harvested from the fjords of Chilean Tierra del Fuego, and they
had a monosyllabic verb that meant 'to unexpectedly come across
something hard when eating something soft' – like a pearl in an
oyster. (English has a monosyllable for that too. Ow!) The magazine story I was writing, long after the last Yaghan swallowed
the final oyster, concerned one of greatest land buys in history.
In the 1960s, Doug Tompkins, an American entrepreneur who
went on to found North Face outdoor wear (and much else),
fell for the cadmium mists and purple grassland of Chacabuco
Valley, which lies in the transition between southern beech forest
and Patagonian steppe. He was there training for ski races. The
Belgian owners of the overgrazed Chacabuco sheep ranch were
not turning a profit, so Tompkins bought their land. Since then
his foundation has acquired millions of acres in both Argentinian
and Chilean Patagonia, turning over most of it, protected, to the
nation. Tompkins expressed the desire 'to explore new ways
of being in this landscape'. There was ambitious talk of creating parklands, restoring biodiversity and promoting ecological

---

1. The ratio of people to square kilometre in Chilean Patagonia is 1:1. In the UK, it's
281:1; in the US, 36:1.

agriculture, of a rewilding programme. For once, action had effectively followed words. In 2015 Tompkins was kayaking on Lago
General Carrera, tipped out, and died of hypothermia. He was
seventy-two. Interest in the Tompkins Foundation's achievement
in Chile lies in its singularity. How many counter-stories could
you and I both tell about the white man's destruction of land?

So I was back on the Carretera Austral, the only road there
is. (I see, going through my cuttings, that I regularly describe
these über-roads as 'fabled', a comment on the role they play in
my storied imagination.) The Southern Highway unravels 1,200
kilometres from Puerto Montt to Villa O'Higgins. It changes its
name at the northern terminus, but basically you could keep rolling till you reached Vancouver. Around Mallin Colorado it turns
into a dirt track, and between Coyhaique and Puyuhuapi into the
snowy, one-lane Paso Queulat. I was frightened there, with no
phone signal and no hope of a rescuer if I slid into a drift. I tried
to summon images of the friendly curtain department. Around
Chaitén, the Pacific appeared. In the pass south of Balmaceda I
saw shaggy huemuls, the endangered South Andean deer. The
natural beauty was an antidote of sorts to the solitude of the open
road, and to everything.

The Tompkins Foundation had fought the usual enemies –
concern that conservation efforts might eliminate traditional,
in this case gaucho, culture, and that predators might increase,
in this case the puma population. By that time I had listened to
argument around the conflict between the human community
and the animal habitat the world over, including the mantra of
species reintroduction, 'There is never a stable line in nature.'
Reintroduction had seemed new in 1999 when I covered a story
about lynx and watched a cautious cat step out of a cage into
deep Colorado snow. An American ranger I met there had worked
on bison reintroduction. His team had transported the two-
thousand-pound animals down a wilderness corridor three states
away, and let them out. 'They beat us home,' the ranger told me.

In Patagonia a fresh puma population had not unleashed terror. Guanaco mating season was approaching, and males were chasing one another, careering over the grasslands to nip a potential rival, or if possible bite off its testicles. North of Parque Patagonia the road entered a temperate zone in which ferns predominated and Chilean dolphins fluked in cold fjords. I saw a lot of salmon farms. Salmon! The growth of pisciculture marked a major shift in the Chilean economy. The business had already taken off in the early nineties, but the industry, mostly foreign-owned, had not swelled into the behemoth employer and polluter it is today. Chile is the world's second largest producer of farmed salmon after Norway. I was shocked to witness the explosion. Was no fjord unmolested? Was no *ceviche salmón* free of antibiotics? Yet firms plan further expansion. Besides that, this time I saw, everywhere, graffiti protesting against mines and hydroelectric dams. The emotional bandwidth didn't exist in 1991. The aim of Chilean youth then was to kick out the evil old. The past lay close, like a roadblock, inhibiting progress in any direction except one that led away from the past. I had grown up during the Cold War, certain that dullard politicians were leading us into another global conflict, and that this time it would be the last one, because it would leave us all dead. The Soviet Union collapsed just before I went to Chile but the legacy of the Cold War was not euphoric release. Now we are the dullards leading the youth to their doom.

Further north, Parque Pumalín (pronounced *Pumaleen*), consists of almost a million acres of temperate evergreen rainforest spiked with mountains and glaciers. It was raining. That's the trouble with rainforest. But it's not like the Amazon rainforest. Pumalín is an attenuated landscape with just twenty species of trees and ten of mammals. Pablo Neruda, one of Chile's Nobel laureates, came from a place just north of Patagonia, and wrote of 'the great southern rain, coming down like a waterfall from the Pole'. Of Patagonian trees, he said 'from their cold green eyes sixty tears splash down on my face ... Anyone who hasn't been in the Chilean forest doesn't know this planet.'[1]

More or less on a whim, after two months in California, I once wondered if it were possible, in the long finger of Baja, to catch a glimpse of the Old Mexico, a land free of casinos, drug wars and the resorts for which Baja is infamous. So I took an Uber from San Diego to the Otay crossing point, walked over the border in twenty minutes (they don't mind you going across in that direction) and picked up a hire car at Tijuana Airport. I drove fast down Highway One, basically a continuation of the (fabled) Pacific Coast Highway, exiting *La Frontera*, the 112-kilometre border zone as soon as possible and stopping only for a two-dollar *pata de mula*, a giant clam served raw in the shell mixed with peppers and chillies. Mexico started for me in the juice of that bivalve.

An inland ranch had caught my eye on the satnav. The turning was just after San Telmo, a village where a man in a kiosk sold individual cigarettes. Driving east into the chaparral with the setting sun behind me, a turkey vulture circled, and the scent of Mexican elderberry drifted through the vents. After an hour I saw a sign to Rancho Meling (it displayed a pictogram of a lavatory) and bumped down a track just as the candelabra cactuses were opening pale green flowers. The forty-square-kilometre Rancho Meling lay at the foot of the Sierra de San Pedro Martír. A handbell clanged for breakfast, lunch and dinner, summoning cattlemen from field to kitchen. I was the sole guest. The gauchos, all young, smiled shyly but otherwise kept to themselves, rising from table to pick out their sombreros and slouch off to make hay even though the sun showed no indication that it planned to cease

1. The first Chilean Nobel laureate was the poet Gabriela Mistral (1889–1957). In 1945 she became the first South American to be honoured in the Literature category (not the first woman – the actual first). Mistral, whose real name was Lucila Godoy Alcayaga, grew up in Montegrande, in the valley where I met José.

shining. The generator hummed for half the day only and there
was no menu in the refectory-style dining room. This was inland
Baja California, a region unrecognisable from the coastal strip,
and one where sombreros on a table linked the land to its past.
The journey to Meling was as much historical as geographical; a
voyage to a place where the old west never died. Proprietor Ari
Meling was the great-granddaughter of Salve Meling, a Norwegian
pioneer who settled at the turn of the last century. He married a
gold panner's daughter, found this fertile valley floor and raised
cattle, bringing workers up from the coast. Prospectors started to
drop by in the 1920s, and the ranch took on a dual role as farm
and simple hotel. The main buildings were constructed of roughly
hewn stone in classic hunting lodge design with red ceramic tile
roofs held up by oak beams, the overall effect a mix of benign
neglect and artless design. Ari Meling had assigned secondary
uses to expired 1942 Chevvies – as tool stores, for example. Quail
skittered across a scrubby hinterland of manzanita and chaparral
ash, and someone had trained one of the dogs to head a football.

The 3,096-metre Picacho del Diablo presides over the ranch
and an eastern desert as dry as the Empty Quarter. Miles of sugar
and lodgepole pines, white fir, incense cedar and the rare San
Pedro Martír cypress abut hills and mesas. From certain points
you see both the Sea of Cortez and the Pacific. Dozing in the
sun, I thought not of stout Cortez, because I was hungry, and the
animal emotions dictate the imagination. The picture in my mind
came from the pages of Sybille Bedford (1911–2006), one of my
top writers on Mexico in English. *A Visit to Don Otavio*, originally
published as *The Sudden View*, tells the story of an eight-month
sojourn in Mexico in the late forties. The book is a confection
of close observation, history, specificity, invention and humour,
together forming the priceless yeast that makes a travel book rise.
That afternoon under the Picacho del Diablo I saw what Bedford
saw, or said she did: a hand holding up to an open train window
a single round white cheese set on a leaf.

Some years later I was in Mexico City to research a magazine piece on Frida Kahlo. She was trending in the UK, largely due to a V&A show. (I won a lookalike competition.) On an evening off from Mexicayotl and 'unflinching depiction of the female experience', I went with my fixer for an evening of *lucha libre*, otherwise known as freestyle wrestling. *Lucha* is the second most popular spectator sport in Mexico, after football. The 16,500-seat Arena México in Colonia Doctores stages professional wrestling shows three times a week, each lasting upwards of four hours. I remember my night at the Arena more clearly than Diego's murals or Frida's Blue House. In the first minute after I sat down, a bulbously muscular man (Clive James's walnuts in a condom) wearing a pair of teeny scarlet trunks bounced off the ropes, wrenched his opponent to the floor and leapt from the ring to a spot on the sticky floor inches from my seat. There he preened to tumultuous applause, arms raised and nostrils flaring. I could see the hairs inside them. The stadium was full. Dry ice swirled, music thumped. The foyer smelled of popcorn, with people milling in and out, talking on phones, taking a break on the balmy street. Inside, skimpily clad female cheerleaders twirled. Unreconstructed, yes, but it was telling, in terms of Mexican social conservatism, that most dancers came from Central America, not Mexico. Dragon Lee, he of the scarlet trunks, also sported a red mask and over-the-knee leggings and boots. (Note, with talk of masks, that this was several years before Covid.) *Lucha libre* is deeply ritualised, and masks – *mascaras* – are a key part of the show. Some of the most famous *luchadores* wear them in public – a friend saw one with an alien-style mask in a restaurant in fashionable Condesa. The grey-haired referee, a figure with his own fan base known as Tirantes, which means braces, was indeed wearing white braces over a black-and-white striped shirt. He threw his arms around and barked decisions to considerable fanfare – he was almost as theatrical as the wrestlers. *Luchadores* can win by pinning the opponent down to a count of three, or by knocking him clean out of the ring, usually for a count of twenty (rules vary according to leagues and weight categories).

Freestyle wrestling landed permanently in Mexico at the beginning of the early twentieth century, loosely adapted from the Greco-Roman sport, the latter a serious business involving oiled limbs, no showboating, and certainly no masks. But this was Mexico. In the arena the crowd grew wild to the point of hysteria. On the red-and-blue plastic seats families stamped and cheered and shouted, some with babies strapped on their chests and toddlers straddling their shoulders. Others wore masks themselves. Grown men tooted horns and jumped to their feet when Dragon performed a particularly spectacular leap from the top of the ropes into the ring. His opponent, *El Kaiser del Infierno*, wore a natty black vest get-up with red horns on his mask. Later in the evening a contestant stepped into the ring wearing a full bandage mask.

And so it went on until Tirantes declared Kaiser the winner. The victor gurned and adopted poses in front of a half-dozen news photographers. Waiters in bow ties patrolled selling drinks and snacks. Coca-Cola developed the Blue Demon Full Throttle energy drink named after *luchador* Blue Demon Jr, who duly became Mexican spokesperson for the beverage. Throughout the night, waves of competitors from the light heavyweight division (*peso semicompleto*) strode into the ring, each man introduced by a figure in a suit. My fixer and I arrived at 9.30, and the show had already been on for two hours. If you miss a bout, the national television channel screens a *lucha libre* catch-up every Sunday morning. My ticket cost 225 pesos, which is £8.50 ($11.38), a hefty sum in a country in which the average family at that time earned £629 ($843) a month. Tickets for big bouts can be much more. 'It's a priority,' my fixer told me. 'For lots of families the *lucha* is their major source of entertainment outside the home, where they watch soaps day and night. There won't be any holidays, but they can go out once a month with all the family to shout themselves insensate at a *lucha*.'

It looked staged all right, and why not? In Britain the TV-screened bouts of the 1970s all featured – or so it seemed – Jackie Pallo. My dad had a copy of his autobiography, which was called *You Grunt, I'll Groan*.

Martha Gellhorn spent four years in Mexico's Cuernavaca, writing of 'billiard table' lawns, Popocatepetl volcano and radiant morning glories. She found the Old Mexico, as I had found it at Rancho Meling, and, released from what she called 'the gilded sewer' of New York, sat in the garden of her ivory bungalow wearing just a hat. It was, she decided later, her golden (as opposed to gilded) age. All that changed in the 1980s and early 1990s when her war reporting from Central America framed the region in tragedy. Panama, El Salvador and Nicaragua especially emerge as pitifully benighted lands squeezed under bulky Mexico and its own hulking northern neighbour. The Sandinistas were in power when Gellhorn was in Managua, and 'For the first time, the state is out of the murder business.' In a poor area of San Salvador she talked to the three sons of Roberto Martelli, a thirty-five-year-old doctor who had worked in maternity services among the poorest women. They, who had seen so many sisters die in childbirth because of a lack of medical care, 'revered him'. He had been a founder of the Human Rights Commission, and it was this work which caused him *to be disappeared*. 'If you think here, you can die,' a staff member of Christian Legal Aid told Gellhorn. Her reports from El Salvador, a member of the UN, made me lose faith in that institution, and over the course of my travels it was never restored. I have only been to Central America twice. One visit provided another antidote, this time to Gellhorn's reality. It reveals a unique brush with Hollywood in my travelling career. When I say 'brush' – there was no direct contact, but I did lie in a superstar's hammock.

If you believe the legend, the green shadows of the Maya Mountains reminded Francis Ford Coppola of the Philippine jungle where he filmed *Apocalypse Now*. Knowing the story of that shoot (in turn hallucinogenic, infernal and miraculous), it's difficult to see why Coppola wanted to be reminded of it. But something about the Belize peaks captured his imagination. When he found an abandoned hunting lodge on the remote slopes of a pine forest, he jumped into the waterfall alongside it and thought, *I could write here*. He lost his glasses in the spume, but kept hold of his dream.

Nestling in the crooked eastern coastline in the lee of Mexico, the marine territory of Belize shelters the longest barrier reef in the Americas as well as coral atolls and a chain of cayes. Formerly a British colony, Belize did not gain full independence until 1981. It remains the only country in the Americas where I keep my British status as quiet as possible. At about that time, Coppola, who as a film-maker was deeply invested in the potential of new technology, had an idea for the fledgling nation. Spotting possibilities in the high literacy rate, the relatively stable democracy and the use of English as official language, he worked out a scheme in which Belize would transform itself into an international telecommunications hub. Coppola approached politicians in Belmopan offering to help them apply for a satellite licence. But it was not the time for foreigners to muscle in on Belize, and anyway nobody in government had the vision to back the scheme. Coppola cast around for other ideas. Perhaps, though, the time has come for a counterfactual screenplay, with Belize Inc. in the Verizon role. *What are they gonna say about him? What are they gonna say? That he was a kind man? That he was a wise man? That he had plans? That he had wisdom?*

Following a barman's tip, the director and his wife Eleanor drove up to the Mountain Pine Ridge Forest Reserve in the western Cayo District. Hiking among Caribbean pines and subtropical ravine flora, they came across an abandoned building on a bank above Privassion Creek. Once a shooting lodge, Blancaneaux had lately been a popular drinking spot with British squaddies. Responding to the romance of the scene, Coppola immediately recognised the paradise hideaway he had sought. He bought the lodge, along with ten acres of forest, for $65,000. Most people would have been intimidated by an absence of infrastructure. But you don't win five Oscars without application. 'It's like being on location for a movie,' explains Coppola. 'You just bring everything with you or build it yourself.' He hired a Mexican architect and set about making Blancaneaux a family retreat. 'I wanted to create a paradise where people could go to enjoy the tranquillity of their surroundings,' he says. 'Someplace far enough away, but close enough.' Challenges piled higher than Caribbean pines. Coppola likes to tell the following joke: 'How do you make a small fortune in Belize?' 'Bring down a large one.' The extended family enjoyed Blancaneaux as a private residence until 1993. Then, after everyone had flown down for Francis's fifty-fourth birthday party, he opened his hideaway to the public. He knew that small is everything in a high-end retreat, so limited construction to seven villas, all serviced by a restaurant and bar in the main lodge building. On my way to Blancaneaux – I travelled overland from Guatemala – I had driven through San Ignacio, the nearest town, on market day. In the main square, a Mennonite family were unloading eggs from the back of a horse-drawn buggy. From there, after an hour's drive up an unmade road, I reached Pine Ridge. In the main lodge a small barman and a large rabbit sat under the fan from *Apocalypse Now*. The former offered me a reviving lime cocktail while the latter eyed me with suspicion. Shortly Coppola's Scottish general manager appeared in the bar. When I had finished the cocktail she led me down several long flights of steep stone steps and ushered me to Coppola's own villa.

Latin American architecture, like Latin American prose, enjoys a command of the exotic and the overstated. A hand-woven

thatched roof tumbled to a generous deck, and in two cuneiform
bedrooms on the flanks of the house, built-in beds had been
simply fashioned from tropical hardwoods. In the bathrooms a
voluminous triangular bath sank into the floor below a roof open
to the mackerel skies of the jungle. In the middle of the villa, the
living room gave directly onto the deck and its fourteen-seater
table. To penetrate the wilderness yawning above, below and on
either side, I hired a horse and guide and headed to a swimming
hole at the bottom of a hundred-foot waterfall. After Horace tells
his friend Bullatius in the Epistles that those who run across the
sea change their sky not their mind, he continues, *Quod petis,
hic est* (What you seek is here). Whatever I was seeking seemed
to be there, uncertainty about the horse notwithstanding. Deep
contentment filled me up like water going into a flask. My guide,
a nubile, smooth-skinned individual wearing a bandana, left his
own steed at the water's edge and bounded up to a ledge fifty feet
over the swimming hole. Perhaps he knew, like Roxanne Sarrault,
that you can never step into the same river twice. His dive was
perfect. It was Hollywood in the jungle – *one side that loves and
one side that kills.*

7

Thin Paths

China

Perched on a natural citadel overlooking the Yangtze, and close to Himalayan foothills, Baoshan has no road in or out, so I walked the last leg. The mud-brick guesthouse, Mr Moo's, had a squatter loo, no coffee and a murderously loud rooster which lived in a permanent dawn. But for three days I perched myself over a bend in the river, watching it change from milky coffee to silver at a thousand-year-old trading post on the Tea Horse Road (*chama dao*), where caravans once carried tea west to Tibet and India, and galloped horses they had bought back to China.

Baoshan is in the top right-hand corner of Yunnan, the province nestling in a dent between Burma, Vietnam and Laos on China's southwest frontier. In the shop (there was only one), cash was stored in a washing-up bowl, and behind the counter, a poster advertising China Mobile had peeled away to reveal a faded slogan from the Cultural Revolution: *ZAO FAN YOU LI* – TO REBEL IS JUSTIFIED. Palimpsest, for once, was not a metaphor. Mr Moo's abutted an ancient entrance gate where people smoked pipes and played cards. Like everyone in Baoshan, they were Naxi (pronounced *Nashi*), members of a 250,000-strong ethnic group of nature-worshippers using the only extant pictographic script in the world. The older among them had never learned

Mandarin. One, Mrs Lao, a sprightly seventy-five-year-old with a face so wrinkled it was smooth, guided me along thin paths down to the Yangtze and through purple Himalayan saxifrage on the mountainsides beyond. Every ten minutes, the Katy Perry ringtone on her mobile went off. *Quod petis, hic est.*

I am, as a writer, a generalist; it's unfashionable, but someone has to be one. A generalist is doomed to skim the surface. I never felt this more acutely than in China. It was not just its size. Or even its history. It was also its fragmentation; its ethnic mix; its geopolitical function and intention. I recently heard a historian say this: 'The fifteenth century was Dutch. The sixteenth Portuguese. The seventeenth Spanish. The eighteenth French. The nineteenth British. The twentieth American.' You know the next one. As a result, geopolitics was in the forefront of my mind – I was looking for the future, not the past. I had been to China a couple of times on short assignments, and in my early fifties set off for two months, focusing on Yunnan and the Qinghai–Gansu–Inner Mongolia arc. At first I felt like Captain Willard in *Apocalypse Now*: 'When I was here, I wanted to be there.' But that faded when the heaped immensity of 'China' took over.

Good intentions faded too when everywhere I saw ways in which the past weighed on individual lives. In Huang San on the outskirts of Lijiang, the ancient Naxi capital, I took a translator to interview a *dongba* – a Naxi shaman (the animistic religion evolved from Tibetan Bon Buddhism). The forty-year-old He Kai sported a Dalí moustache and oiled hair and wore a gold waistcoat over scarlet robes. We sat round a fire in his single room while he tugged on a metre-long pipe with a cigarette poked in the end. Paintings of animal spirits danced in the firelight weaving on the wall. Mr He was a sixth-generation dongba. I asked what his role involved.

'I speak to the spirits on behalf of the people. I give advice on failing harvests and conduct ceremonies – weddings, funerals,

prayers for convalescence, that kind of thing. At festivals I pray
for good fortune.' He had never been to school. 'As a child I used
to go around the mountain villages with my grandpa, who was
also the medicine man.' He had to beg for several years while
learning the Naxi script, which only the priestly caste writes
and reads. I asked if he would write me a line, and he brought

out a bundle of the insect-resistant paper Naxi have made for generations and started writing – painting rather, as he used a brush. 'Red Guards destroyed most of our books in the Cultural Revolution,' he said as sweeping hieroglyphics crept across the paper, 'and either killed dongbas or compelled them to abandon their profession.' Naxi had never really recovered from the murderous horrors of the Cultural Revolution. Was there even to be a future? 'I worry,' said Mr He. 'There are only thirty dongbas left. In village classrooms, Naxi children have to speak Mandarin.' Beijing boasts about 'one big family', but the goddess of democracy statue hoisted in Tian'anmen in 1989 has yet to be made flesh.

I longed for female company in Yunnan, and found it (eventually) in the rural baths. I bathed with Naxi women who cleaned one another's backs with a washing-up brush and unravelled topknots to soap long black hair. When they had finished, they rinsed their knickers in the same water. They pointed at the best place to stand to feel the impact of the flowing spring, and stared at my DD poitrine, never having seen such enormous appendages. When they felt more confident and sufficient smiles had been exchanged, they asked if they could touch them. The roofs were arches of dirty glass, and afternoon sun slanted onto cobwebs and cracked plaster, and onto the girls' spent shampoo sachets floating on the surface of the sudsy water.

Near the end of the journey from Shangri-La to Xanadu, a burning wind whirled in from the pamirs. On the world's highest dune, the top crust of sand shattered underfoot, like thin sea ice. Inner Mongolia's Badain Jaran desert, as far to China's north as Yunnan is to its south, extends over fifty thousand square kilometres and like tiny Baoshan has no roads. Climbing the big dune – the one Mongols and Han alike call 'the Everest of the desert' – turned out to be a slog. From the top it took ten minutes to slalom all the way down and plunge into a turquoise lake where the salty water worked up a lather when I found my soap. I travelled in a jeep across the Badain Jaran with a fearless Mongol driver called Alat. Shortly after leaving the tarmac for a week, he chose a picnic spot next to a two-metre stone carving of the head and hands of Genghis Khan. There under a cornflower sky we ate yak jerky, heavy buckwheat bread, hard-boiled eggs, tomatoes, watermelon and peaches. Winds that ferried in the tang of the Gobi ridged certain slopes, and on the summits, the same winds whipped up plumes of sand, turning dunes into smoking volcanoes. Alat and I climbed many of these sand mountains, always discovering a lake on the other side shining like an eye. Uniquely for such a parched environment, more than 140 lakes dot and dash across Badain Jaran (the name means 'Mysterious Lakes' in Mongolian). Near one, we pitched up at a guesthouse consisting of two facing rows of mud-rooms. There were no bathrooms, just a rainwater butt with scoops and basins, and the private dunes. A juvenile camel yet to grow humps wandered around. When it became too dark to read, Alat taught me how to text in Mandarin (a fiendish operation involving the Pinyin transliteration system). In the middle of the night, the dunes boomed as wind pulled the top layer of sand down the slopes – what Marco Polo called 'the tramp and clash of great cavalcades at night'.

The owner of the guesthouse, An Fu Jiu, was a sixty-year-old Han. He was short and wiry, with a bald patch in otherwise lustrous hair. He wore trousers that were too big for him, secured by a belt. Born in the neighbouring province of Gansu, Mr An had travelled with his parents to the Inner Mongolian desert when he was four. The family farmed, and in the famine caused by baby-eating Mao's Great Leap Forward came close to starvation. 'When we first got to the desert only ethnic Mongolian camel herders were here,' Mr An remembered as we ate his soupy breakfast noodles, 'and my parents looked after their animals. There were no houses then. We stayed in other people's *gers* [yurts].' He had built the guesthouse himself and made the ceilings from rolls of uncut plasticised cardboard packaging of Butterfingers Blasted candy. Tate Modern would have paid millions to take it home strip-by-strip and put it up in Bankside. Mr An said it kept out dust and sand. As he spoke, light varnished the crown of his head. He had a vegetable garden behind reed fences. The way he conjured aubergines, tomatoes, courgettes, onions and much more out of sand was akin to a Confucian miracle. A recently acquired generator meant he could pump the well electrically if he had been into town to buy fuel; a shadouf-like arrangement stood in the middle of the garden for the other times. Many writers have noticed the desert's capacity to simplify consciousness itself and I felt it among the parsed dunes around Mr An's.

The Qing dynasty opened Inner Mongolia to Han in the eighteenth century. (Outer Mongolia won independence in 1921, which is why the map of China looks like it's had a bite taken out of the top.) Genghis Khan's grandson built his summer capital, Shangdu, known in English as Xanadu, in the far east of the province, now a few acres of collapsed mud walls, itinerant sheep and plastic bags cartwheeling in the breeze – more Ozymandias than pleasure dome. Uighurs are the current story, but in the 1980s, China locked up ten times more political prisoners than the Soviet Union and shipped most of them to far-off Gansu and Qinghai. This was called 'Reform through Labour'. Alat and I took Highway 312, formerly known as the Silk Road, out of the desert into corridor-shaped Gansu, where ghostly prisoners marched to oblivion across the scrubby plains. We turned south to the Black River and trekked across escarpments striped with the windborne loam of the Gobi. A lone mushroom collector carried a red bucket on his back, the hump the camel calf had not yet grown. From the top we saw grasslands spread out like an inland sea. At the bottom we took butter tea on the riverbank with a family of Tibetan herders. They had a hundred yaks and seven hundred sheep ('This makes them very rich'). Bricks of yak dung lay drying among primulas next to a queue of solar panels, and in the neat *gers* a picture of the Dalai Lama presided over a shrine. We sat on sheepskins in the sunshine, breathing in dung smoke while a ten year old translated her mother's Tibetan into Mandarin and a chained mastiff expressed a longing to eviscerate us all. At dawn the next day we crossed into Qinghai, formerly Amdo, one of old Tibet's three provinces (and about the size of Turkey). A party of gold-toothed horse-traders were already up at their tented camp. They had decked their steeds in silk sashes and were trotting them in circles. As the sun lifted above the Qilian Shan, marmots crept out of their burrows. Later, we ate yak yoghurt in a place where vultures swooped so low you could see individual feathers. Mountains and grasslands receded in every direction. And it went on. Lake Qinghai, the biggest in China (4,500 square kilometres), shimmered under a lacquered sky. I might have been the only person in the country. The other 1.3 billion lived in another China.

One night, after a week in the dunes, I drove into Xining, capital of Qinghai. Banks of neon blinked. A third of the population is Muslim, mostly Hui, Uighur or Salar, and eighty mosques shoulder up to galaxies of fluorescence – at ten in the evening people and traffic jammed the streets. In the morning stallholders in the Muslim Quarter touted chicken feet, unshapely boulders of yak butter and pyramids of tea, while liver nodes lay on slabs like murdered Roman emperors, vendors batting away horseflies with plastic paddles. It was August, and a rubbish truck played 'Jingle Bells' over its loudspeaker.

At Tongren in southeast Qinghai, a dentistry stall displayed a range of pliers alongside magazine cut-outs of Tom Cruise flashing his gnashers. Tongren is a monastery town, and monks with good teeth hurried by. Around the temples and lamaseries of the fourteenth-century Longwu complex the air smelled of bitter oranges. In the kitchen, old women with faces like walnuts twisted cotton into wicks and others ferried kettles of melted yak butter. At the sound of a gong, monks disgorged from alleyways, pulling cloaks tight. They carried in their hands, or resting on their shoulders, yellow headdresses shaped like Mohican haircuts (Longwu belongs to the Dalai Lama's Yellow Hat sect). Three hundred resident monks entered the courtyard for the day's *tso pa*. Before the ritual, in which monks test one another on Buddhist philosophy (the pious can recite a thousand pages by heart), everyone sat chanting, cross-legged on cushions, while a lama intoned in a froggy baritone. The junior monks – the youngest was six – whispered and tittered, someone threw a note to the lama screwed up in a ball, and a couple of greybeards nodded off. After half an hour, another gong started the *tso pa*, a performance enacted first in pairs then in groups of twenty. The effect in the dying light was balletic.

Beijing officials criticise westerners for romanticising Tibet. Yet when it suits them, they do it. Yunnan's Zhongdian allegedly inspired James Hilton's 1933 novel *Lost Horizon*, a utopian fantasy of Shangri-La, a paradise at the foot of the Himalaya (in fact, Hilton had never been anywhere near Yunnan). In 2001, scenting money in the rising popularity of Hilton's story and the millennial quest for paradise, the government renamed Zhongdian Shangri-La and turned the town into a tourist resort where hundreds of thousands of Han from the eastern seaboard indulge in the Tibet-worship Beijing claims to despise. Further evidence of governmental capacity to alter the narrative appeared at Labrang Tashi Khyil in Gansu, founded in 1709 near the lip of the Tibetan plateau on the Daxia, a tributary of the Yellow River. Sitting dead in the middle of China, in a perfect bowl shaped by ridgebacks of the Dragon and Phoenix mountains, for centuries Labrang formed the centre of a web of trading routes where Han, Hui, Tibetans and Mongolians gathered in the lanes. A monk took a picture of me on his mobile phone. To him, Labrang belongs to Tibet proper, and its founder was the first Living Buddha. In 1958, during a period of especially savage Han repression, the monastery closed for twelve years. Now pilgrims walk around the perimeter in sheepskin boots, spinning prayer wheels. A pro-Tibet farmer had recently set himself on fire in a Labrang courtyard. The monks prevented police from retrieving the corpse (it remained in the monastery at the time of my visit). 'In order to realise their separatist goals,' quacked Xinhua, the official news agency 1,300 kilometres away in Beijing, 'the Dalai clique has incited some people to self-immolate. This is despicable and should be condemned.' In fact, the Dalai Lama opposes self-immolation.

I wondered, given the omniscient and omnipotent presence of a malign government, what people left out of their accounts when they spoke to me. An Fu Jiu with his home-grown tomatoes just missed what he called 'the ten years of chaos' – the Cultural Revolution – and the girls in the sudsy Yunnan baths had certainly escaped it. But they knew. The Naxi dongba carried the knowledge like a humped burden on his back: if he hadn't experienced the worst of it, his people had. You could see the shadows of Mandarin characters between the lines of his Naxi pictograms. Mrs Lao with the Katy Perry ringtone, the dentistry artist at Tongren, the rich herders, the monks – the past wasn't dead for them; as Faulkner said, it wasn't even past. In *Chinese Lives*, Zhang Xinxin and her colleague Sang Ye's oral history, Li Xiaochang, born in Beijing in 1949, the year the People's Republic was founded, spoke of her relocation not far from An Fu Jiu's tomato patch in Inner Mongolia. Earlier, in middle school she had rampaged with other Red Guarders beating up anyone who wasn't rampaging with them. 'Even today,' she said, 'I don't know who some of our targets were,' though she did know the Nine Dragons and a Phoenix gang, whom she and her colleagues brutalised 'till they begged us Red Guard ladies for mercy'. Xiaochang went on long train journeys 'to exchange revolutionary experience'. The carriages were so crowded that young Red Guards sat in the luggage rack like my peanut-picking friend in Alabama, but in Hebei and Tianjin in the 1960s 'the racks couldn't hold the weight and they'd fall down'. At first Xiaochang enjoyed Inner Mongolia. She learned to ride a horse and to drink tea with milk. 'If now I have a sense of wonder,' she told Xinxin and Ye, 'or of powerful, indefinable nostalgia, if I have an understanding for melancholy and quietness, it all goes back to those days.' This was *Heimat*. But Inner Mongolia turned sour when her herdsman host raped her ('the whole family was there and they helped him'). At the time of the oral history interview she was working for the government on 'archival materials about ancient buildings', struggling, as a mature student, to acquire the university qualification she needed to guarantee job security. It was 1986. She said then, looking to the future, 'Our country's got to develop.'

I glimpsed the future myself in Shanghai eleven years after Xiaochang expressed her urgent wish for China. It was 2 May 1997. I was heading to Lanzhou and beyond on a magazine assignment and staying overnight in a business hotel with a photographer; we had met for the first time at Heathrow. I had just flown in, and couldn't sleep; I watched the first pink streak of the new day bleed from the ocean horizon. The hotel room had BBC News. I can see Shanghai waking up beneath the seventeenth-floor window, the small branch of blossom on the pile of fluffy towels, and Tony Blair raising his hand to ecstatic crowds in the Royal Festival Hall. That night, in another hotel in Xiahe County, the photographer and I heard on the World Service that we had voted in 120 women MPs, and that Blair had put an unprecedented five women in his first Cabinet. A new dawn.

As for the practicalities of travel in China: squatter lavs are de rigueur in stations outside metropolitan areas (and often within them – the gleaming new Kunming Airport had no sit-downs) and I broke fresh personal ground at an open-air bus station in Guangxi when I crouched next to a stranger in a two-at-a-time stall. Patrons enter in pairs as they did in the ark and squat companionably over adjacent holes. Catching a bus from Lijiang to Zhaotong had involved a search for the vehicle's *number plate* rather than its *number*, as Yunnan buses don't have numbers, and this at a crossroads marking the starting point of a hundred buses. On the four-hour minibus trip from Lijiang to Baoshan, the driver chain-smoked and talked on his phone, even on the hairpin bends. Later I sat in the aisle for the whole of a six-hour public bus ride to Lake Lugu. On a two-hour journey to Jinshanling in Hebei Province northeast of the capital my private car passed three accidents, one of which had resulted in a pair of corpses covered in their coats, now shrouds, on the tarmac. But my love affair with trains found its apotheosis on Chinese fire-wagons (*huoche*). The old-fashioned Ks which trundled me from Jinchang to Xi'an in thirteen hours; the Z-class overnight express which took seventeen hours to get from Lanzhou to Beijing; the soft sleeper (*ruan wo*) on night trains far from Auden's postal orders. In the 'hard sleeper' carriages – the ones with seats and no bunks – people milled up and down spitting sunflower seed pods or smoking at the end of the carriages, babies cried, small boys jabbed at electronic games, families played cards and many slept open-mouthed as brown hills sped by. I liked the ferocious conductresses who prowled up and down in peacock-blue shirts with epaulets, black trousers and preposterously huge black-peaked caps with a yellow band. Somewhere in Shaanxi, three construction workers ate noodles in the restaurant car and swigged from green bottles of 56 per cent proof rice wine, and at the end of the meal they sent a bottle over to my table.

The long train journeys revealed swathes of heavy industry eructating dense black smoke; plastics factories that made the air inside the carriage smell like Superglue; immense fleets of coal trucks covered in red tarps lumbering everywhere. In Shaanxi the sky was permanently dark. No wonder Chinese call that region the Black Belt. Yet an engineer told me that his hometown Chongqing was worse. 'We call it Smog City,' he said. Chongqing was the poster boy of China's dash to modernise.

Having split from the province of Sichuan in 1997, it became an autonomous zone with a still-exploding population of 31 million. The engineer had seen the sun five times in two years. Pollution and environmental depredation reached Inner Mongolia late, but they had come. To a certain extent they arrived as a result of publicised contamination in the populated industrial regions of the south. Just as Beijing relocated people by the tens of thousands to make way for gargantuan engineering schemes, it had relocated factories so visitors elsewhere saw blue skies. Push the crap out of sight! Local party bosses had poor eyesight when it came to the stringent environmental regulations sometimes adhered to in the south. As the twenty-first century rolled in, Inner Mongolia rivalled both Shaanxi and Shanxi as the country's biggest coal producer; you can see the mines on Google Earth. When I was there, the province was pumping out the highest per capita carbon emissions in China. The fragile grass ecosystems had deteriorated and the province had become the main source of dust and salt-alkaline storms that surge over the northeast of the continent every spring and even infiltrate the western press, generally indifferent to Asian pollution. Environmentalists spoke of 'a calamitous loss of arable land' in Inner Mongolia, the calamity presumably occurring to the food supply chain. The city of Ordos where Genghis Khan once pitched his six white yurts now mines a sixth of China's coal. Grotesquely, in August 2012, the Chinese authorities hosted the Miss World contest in Ordos to raise the profile of that already super-polluted city and attract inward investment for yet more industrial development.

Plenty of interdisciplinary experts maintain that the main environmental threat to China, ahead of coal extraction, is water exploitation. China has built eighty-seven thousand dams in under a century, many close to seismic fault lines. In northern desert regions like Gansu, according to Jonathan Watts in *When a Billion Chinese Jump*, 'the shortage of water . . . prompted some of the most desperate remedies'. Hundreds of thousands of workers died in the sixties and seventies when poorly constructed dams collapsed, though the outside world knew nothing for a long time, if ever. Mao tried to re-engineer nature to many ends, and his successors are still at it. Watts writes, 'although the leadership have tilted to the green after the insanity of Mao's mountain-moving, governance is weak at local level'. As conservation and anti-pollution laws proliferate, everything gets worse. Large-scale, high-rise construction was under way in every province I visited in China. Entire districts sprouted, yet tens of thousands of tower blocks stood empty. Inflation was running then at 8–10 per cent whatever official figures said, and nobody knows what will happen when the property bubble bursts. In addition, you don't get to be the world's largest energy producer without belching out a lot of crap (not to mention irreversibly poisoning the water table). In Xining, in Xi'an, in Beijing, the traffic was filthy, metaphorically, the pollution filthy, literally. The headlong rush to develop has pushed so far west that it was even approaching my Yunnan paradise. High up in Mr Moo's eyrie, as I watched the Yangtze change colour, I contemplated a gash through the Himalayan yew on the mountainside. It was a dirt road, inching nearer.

Xiaojun Wang, known as Tom, was Communications Director of Greenpeace East Asia, based at that time in Beijing. I had lunch with him at a restaurant opposite his office in Dongcheng. Snake and bullfrog were on the menu but we stuck to tofu, noodles, boiled pork and spicy greens. It turned out that the Mao-to-Market transformation was not just poisoning people with bad air. It was eliminating their culture too. 'China's mad drive for coal,' Tom told me, 'has marginalised formerly nomadic ethnic Mongolians. They're either made to settle in a city, or given a patch of contaminated land – in a recent case we supported, a horse died after drinking from a river.' Users of China's burgeoning microblogging scene (Weibo, as I write a community of about four and a half million) have a lot to say about their politicians' disregard for the natural world. 'Recently,' Tom told me at the end of our meal, 'the government in Hohhot [capital of Inner Mongolia] spent tens of thousands of yuan on sculptures of sheep and horses to adorn the grassland next to some power plants in order to attract tourists. The real animals were long gone, along with their herders. That's a very Chinese solution: just make it *look* like it used to.'

The dunes of Badain Jaran continue to sing their nightly songs, and Mr An still tends his desert vegetables, at least for the time being. The deep peace of nature and the ferocious serenity of the monasteries – both still there in abundance – keep company with the chaos of insane industrialisation. It seems unlikely that this particular yin and yang can retain any semblance of balance. Jonathan Watts, the author and foreign correspondent I mentioned earlier, concludes however by highlighting his main concern for the Chinese environment. It is not water supply or coal extraction, but rampant and unbridled 'Shanghai consumerism'. If it continues, he writes, China will have to produce millions more cars and air-conditioning units every year to satisfy the needs of an emerging middle class. At about this stage of the travelling life described in these pages I knew the end of the environmental story, or thought I did. I knew what the conclusion will be, though it will not come in my lifetime. Planet Earth will survive, as it always has, at least in the billion-year time frame. We as a species will be long gone. I was beginning to see the human race as loveable, as it was so pitiable. My position was not one of gloom. It was of a kind of wearied acceptance, a facet, I see looking back, of the post-fifty retreat from Moscow. I did still feel angry all the time about almost everything, which may seem like a contradiction; perhaps fires blazed in the background. Withdrawal, yes, though not capitulation. I do remember though that we still had a landline at home in my Chinese-dominated year, and when it rang I was glad it wasn't for me.

As for the fire-wagons, they were political, like everything else.

Back in 1973 Mao announced construction of the world's highest railway, a 1,900-kilometre line from Xining, where the rubbish truck had played 'Jingle Bells', to Lhasa. Everyone had thought the Kunlun uncrossable. The outside world complained – this railway was a scheme to make it easier to control and repress Tibetans. That, of course, had been the idea when the British built railways across India. The last leg of track opened in 2006. Fifteen years later the first electrified fire-wagon chugged across the Tibet Autonomous Region on a line, according to Xinhua, 'expected to promote exchanges among people of all ethnic groups, and accelerate the high-quality development of Tibet'.

In the third century BCE, warrior emperor Qin Shi Huang united the feudal princelings of the Middle Kingdom and made a country called China. He did it more than a thousand kilometres southwest of Beijing, in Xi'an (pronounced *See Ann*). During his thirty-six-year rule Qin standardised laws, currency, weights, measures, axle lengths and a written language. He also built the Great Wall, burned any book that did not glorify him and buried hundreds of scholars alive. Xi'an's terracotta warriors didn't do much for me but I liked cycling round the paths on the wide top of the fourteenth-century elevated city walls (intact, uniquely in China). As capital of the Qin empire, Xi'an grew to become the greatest city in the world. It drew scholars and merchants from across Asia and marked the eastern end of the Silk Road, the place from which camel caravans a thousand strong set off through the gates I was cycling over, taking silk, gunpowder and chrysanthemums west and lumbering back a year later with bolts of linen and sacks of cucumbers. People were hurrying home from work over the cobbled road on the top of the walls, some no doubt with a cucumber in their packs. At a tea shop, two men played mah-jong besides long-spouted copper kettles brewing various types of tea, and in the courtyard behind an eight-piece orchestra battled through Qin folk opera tunes as a succession of women clutched a microphone and shouted lugubriously. Across the grid of streets within the walls, blackened tenements squared up to pockmarked offices and the lotus bud finial of a temple. I got off my bike to look down into Renmin (People's) Square. Much of Xi'an's civic architecture went up in the fifties, built with Moscow's money, and the blocky, unforgiving exterior of Renmin was characteristically post-Stalinist – the kind of architecture that says, *Keep Out*. From the west gate, a ray of dying sun flashed off the octagonal roofs of a blue minaret. I recognised the warrens of the Muslim Quarter where I had nibbled cloves of raw garlic at lunchtime in a Hui café that smelled of boiled greens and sesame oil. Beyond the walls, lights flipped on in the windows of a thousand tower blocks. Xi'an is the capital of resource-rich Shaanxi. Monster cranes bristled in every direction. What I took for a light show one evening from my hotel room turned out to be the flashing lamps of welders working through the night on the renovation of a three-hundred-room building. The last tour buses returned from the Terracotta Warriors, Qin's necropolis where eight thousand men and horses wait loyally in battle formation. It took 750,000 workers to create the army: each figure, famously, is different – I noted that even the pattern of studs on the kneeling archers' shoes varied. (I thought of the thousand individually carved soapstone Asebu figures in Esie, Nigeria. Why were they not well known?) Excavation has been in progress in Xi'an for a quarter of a century. When I visited, three workers were carting mud and soil around one of the pits in wheelbarrows. Three-quarters of a million to create it, three to reconstruct it.

I trekked from Yunnan into Sichuan and my horseman yelled the old song of the mountains. A tang of snow wafted from Himalayan foothills. The journey began at Lake Lugu (*Lugu Hu*); specifically, at Lige on its northwestern shore. Stepping across rills of sand ringing the fifty-two-square-kilometre lake, I hopped into one of the shallow-draught rowboats called *zhucao* to reach Liwubi Island, where Mosuo people settled early in the sixteenth century. The forty thousand Mosuo represent one of China's fifty-five recognised ethnic minorities, and almost all of them farm close to Lugu, the women conspicuous in the fields in headgear Winsor & Newton paint tubes would designate Opera Rose. Short and strong, Mosuo resemble the ethnic Tibetans with whom they share kinship. But unlike Tibetans, Mosuo are matrilineal – the last genuine matrilineal society on Earth. They enjoy an unusual conjugal system. Women and men pair up by mutual consent in the female's house for a night, or maybe a year, and when the arrangement breaks down the man returns to his mother's home with no harm done. What kind of fabulous system is that? Offspring remain in maternal care, and the eldest woman in the family takes charge of the household. I enquired about conversion, but it turns out you can't convert into an ethnic type. Han Chinese call the system 'walking marriage'. At any rate, whatever the 'system', women seemed to be doing all the work. I saw them every day, hauling teetering loads of hay on their backs or tugging buckets of water from the well.

In September 2021, half a millennium after the first Mosuo woman pulled up water on Liwubi Island, a British newspaper ran an interview with the Oscar-winning actor Matthew McConaughey. It was one of those features in which a celeb describes his day. 'I need my nine hours sleep,' McConaughey revealed, 'so I might not be awake till nine a.m. Camila needs a lot less, and she'll be already up, getting the kids ready.' This remark, made without shame or self-knowledge, lies in my heart like a bar of lead. The actor went on to say that once he had finally stirred he enjoyed reading a little *Daily Stoic*. One Twitter user posted, with eloquence possible only in a tweet, 'Oh Matthew McConaughey. Mate.' It wasn't about him. It was about everything, and I couldn't stand it. The Mosuo didn't care though, women or men. Beyond Lige, square fields of sorghum and cabbage shrank beneath rounded mountains, clouds suspended like canopies. Wooden homes punctuated teal valley seas, many flying white streamers from their curved eaves to honour ancestral reincarnation. A water buffalo picked out a delicate path. A 5.7-magnitude earthquake had struck the area three months before my visit and a tender-blue tent donated by a domestic NGO stood close to each home (in most cases the quake had not destroyed houses – wood does better than concrete – but the regional government was still advising residents to sleep in tents in case of aftershocks). At the Zhamei Buddhist temple in Yongning, ghee lamps illuminated painted statues of Gemu, the Mosuo mother goddess. The gold-plated copper roof of the main pavilion had survived not just the quake but also the Cultural Revolution (just), but much had not outlived that painful decade. Zhamei means 'No War' in Tibetan.

Houses dotted a river valley between spiny hills, cornfields and mud-brick walls encircling vegetable gardens. This was Wenquan. Himalayan poppies bloomed, along with azalea, alpine primulas and a delicate flower so blue it was vulgar. In the single guesthouse my Mosuo host had painted the walls with images of fish, lotus and peacocks and stacked one-and-a-half-metre-long cured pigskins, their mouths and other orifices sewn. In the mornings, I sat on them to drink butter tea. Although not as famous as the Silk Road, the Tea Horse Road was equally important, and for centuries caravans trekked along the same thin paths that I was following out of Lake Lugu and onwards to Sichuan.

At one end of the Mosuo homes into which I was often invited, a monumental black sow occupied a small room from which she exited and entered with decorum when she needed to use the facilities in the adjoining barn. Yes, they had house-trained hogs. Off the courtyard, a matriarch lounged in the bed of honour next to the fire. Mosuo women are remarkably long-lived, no doubt because they don't have to cohabit with the same man for more than a year. I met one who was ninety-six and another 103.

Outside, sprightly seventy-year-old daughters wielded cleavers among piles of spinach. I walked for three hours one day and met only a goatherd lying next to a fire with an antiquated rifle at his side. When it started to rain, he plucked a leaf and whistled, repeating the procedure over many minutes, one leaf per whistle, and eventually two hundred charges assembled, ambling in from every direction. At the next village a stupa had recently been erected on a hill, and a festival had been declared for its filling and sealing, a sacred ritual in which Mosuo entomb written prayers, ancestral relics and assorted gifts. Soon after dawn one day villagers began carting up plastic bags of hard-boiled eggs, apples, cured ham and bottles of pop. Shortly after eight, six lamas in wine-coloured robes and yellow sneakers started to incant, and the putting-in and cementing-up proceeded for some hours, finishing with a dozen rounds of firecrackers and vigorous blowing into conch shells. At the end of the day, leftover food was divided into takeaway bags. The last rays of the setting sun glittered on the rice fields and peace settled, conferring, or so it seemed to me, a sense of unending continuity with the landscape where Mosuo have lived for unrecorded generations. They have survived everything history has hurled at them, but the worst may be yet to come. Han continue to push west. As I walked, developers were moving cranes and dozers into those secluded valleys to build hydroelectric plants, and Yunnan's forest cover had already halved. Hydroelectric plants might not emit carbon but they attract heavy industry. The remoteness of Yunnan had once protected it. With 4 per cent of Chinese land area, the province is or was home to more than half the country's vertebrates, higher plant species and orchids as well as almost three-quarters of its endangered animals, many of which do not exist elsewhere. It was China's beating green heart, a living fragment of what the country had once been. There was little outward sign of environmental vandalism in that isolated region, but it is there: poisoning rivers, drying out lakes, polluting the thin air. Mosuo, like Naxi, practise nature worship and prioritise *balance* between man and nature, believing Earth will grow angry if abused. And that, as you know, has come to pass.

The terrain was rough on the next leg and the way uncertain, so I travelled with a packhorse and horseman. The latter, Gesang, was an ink-haired Mosuo from Wenquan, and he had learned a little English working as a waiter in Lijiang. His robust pony had a short, neatly clipped mane, and the tinkling of bridle bells signalled our passage. As soon as we entered deep forest, Gesang began collecting wild mushrooms, and the activity seemed to please him; he chuckled and sang to himself like a kettle. At our lunch stop – he told me the spot was called Twenty Turns – we roasted the mushrooms on willow sticks and ate them with cold scrambled egg left over from breakfast, served from a plastic bag. Later that afternoon Gesang recognised a woman carding goats' wool. He stopped to talk with her. The woman was minding a pig and fourteen piglets. After their conversation, she and Gesang sat for a long time listening to the silence of the hills. I fell asleep, and a piglet woke me by licking my face.

That night Gesang, the pony and I crossed into Sichuan and put up at Lijiazui, a half-Mosuo, half-Mongol hamlet, ancestors of the second group having galloped in with Genghis Kahn. Our accommodation, on a mud street, enjoyed cardboard mattresses but no sanitation. Wires from a recently installed electricity system trailed nomadically from bare room to bare room, but no longer worked (had they ever, I wondered?). Yunnan is a poor province compared with Sichuan, but there in the quiet border-lands the difference did not exist and never had. Every Mosuo seems to know every other, and that night Gesang's cousin's wife's family invited us for dinner. I sat cross-legged in the place of honour round the hearth and ate grated potato, pork and pickled persimmon. At midnight, our host led us home holding a flaming torch. The cranes seemed a long way off.

*

A menagerie of beasts like Gesang's pony supplied the motifs of the travel-writing boom of the late seventies and early eighties – I've mentioned Dervla Murphy's Cameroonian mule; Isabella Bird, one of my Victorian girls, rode horses over half the globe. I never mounted Gesang's pony though. I still remember the relief in seconds spent tumbling headlong from a steed on the Chilean island of Chiloé. I am a poor horsewoman, and a fearful one, and a horse senses inadequacy. But often in my career I have been obliged to saddle up. On Chiloé I was with a genial Englishman called Chris Sainsbury (his family once had something to do with the cheese counter, or so he said) and he had procured two nags to get where I wanted to go – a fjord inaccessible by public transport. As my horse's fetlock plunged into a sphagnum bog, and my head followed it, I thought, *I never have to get on a horse again*. And I never have.

Boats, though, are conducive to thought. I find equilibrium in the non-place of a vessel at sea; there I can be a true traveller in residence, like Maeve Brennan. One of my at-sea stories related a trip far from China in 2016, across the South Atlantic on the Royal Mail Ship *St Helena*. The RMS, as she was affectionately known, was the last working vessel of her kind. For generations the ship's three-week cycle between Cape Town and St Helena determined the rhythm of life on one of the remotest inhabited islands in the world like the rise and fall of sap. But an airport was scheduled to open on St Helena the year of my visit (there were delays), and the scrapheap beckoned for the last mailship. Something old-fashioned already clung to the simple white livery of the vessel – a whiff of another age. You half expected the stocky funnel to erupt with steam. On my outbound voyage the purser presided over the South Atlantic Ashes, a cricket match in which nets are erected on the sun deck and a taped-up ball of string pressed into service. Runs are awarded according to the height of the shot. The purser-umpire smoked a cigarillo throughout. Half the 135 passengers were Saints, as St Helena residents are known, many returning from medical treatment in Cape Town. Islanders who have spent their working lives abroad often sail home to die. Some don't make it back, and the tiny hospital on B-Deck becomes a mortuary. Like most of the crew, Captain Andrew Greentree was a Saint. He had begun his career twenty-four years previously as a deck cadet on the predecessor of the current RMS. I spoke to him in his modest suite, where a portrait of the Queen hung in the living area. The skipper worked two and a half months on and the same period off. 'The unusual thing,' he said, 'about this ship – besides the fact that she is half-freight and half-passenger – is that the crew have grown up with one another on that tiny island.' St Helena covers 121 square kilometres. 'We've had three brothers on the same roster. On this voyage we have a mother and daughter.' My cabin steward regretted the forthcoming end of an era. 'We welcome the airport,' she said, 'but it won't be the same.' Like many Saints, the ship has served her distant country. During the Falklands conflict she was seconded to the Ministry of Defence for two years, sailing south as mother ship to a pair of minesweepers. She was ingrained in island history, and many Saints knew instinctively when she would next be anchoring in St James's Bay.

It was not just the airport that put an end to the mailship. Email has too. When Greentree began his career, each voyage carried four hundred sacks of birthday cards, bank statements and Dear John letters, some no doubt an earlier incarnation of the one at the South Pole which communicated the irreducible truth that His is Here. On my trip, just three sacks lay below decks.

On Sunday, the captain held divine service in the A-Deck lounge. Taped organ music accompanied us few as we belted out tunes from a special RMS hymn pamphlet (*Immortal! Invisible!*). Captain Greentree wore all white: crisply pressed shorts, knee-high socks and lace-ups. The ship continued at a stately fifteen knots but rolled a little, so there was no genuflecting, and we steadied ourselves on the back of the chair in front while standing to sing. As the Prayer for Seafarers reached its conclusion, we looked out at the Atlantic and the ghosts of thousands of sailors who had heard that prayer on a windy deck since the first man sighted St Helena in 1502, not knowing what the future held (but who does?). The captain was not a priest, so could not bless us; but peace came down like a benediction in that stuffy lounge.

Thus a seagoing vessel is a writer's gift and waterborne accounts conjuring a metaphorical inner voyage have existed since the *Odyssey*. But so have the bipedal variety. Something about the rhythm of walking suits prose. Taiwan-raised Sanmao, born in Chongqing in 1943 (her birth name was Chen Mao-ping) prefers in her many books to travel at a walking pace because, she says, the speed is suitable for registering one's surroundings, in her case far from home, in the Western Sahara. I had first come across Sanmao's writing in a Beijing hotel room, where I found a copy of *Stories of the Sahara* (1976) under the bed with an empty crisp packet rolled into a bookmark. She was among the first media stars in China some years before the West sunk its tentacles deep – McDonald's did not open its first outlet till 1990 (it was in Shenzhen). Pirated editions of her travel books were everywhere, and on the anniversary of her death a Communist Youth League Weibo post received more than a hundred thousand comments. Her accounts of walking trips adhere to her contemporary, Jonathan Raban's, insistence that 'the travel book, in its deceptive simulation of the journey's fits and starts, rehearses life's own fragmentation'. Raban goes on to say that, 'More even than the novel, it [the travel book] embraces the contingency of things.' Only a person who believes in the primacy of fiction could write 'More *even*'.

Gavin Young, in *Slow Boats to China*, voyaged to Canton in twenty-three vessels over seven months, his ships characters in the story. In that volume and its sequel *Slow Boats Home* (1981 and 1985) Young raises the issue of why the 1930s and 1970s were the golden ages of travel writing but does not question why so few women were in on the game. As twice chair of the Stanford Dolman Travel Book of the Year Award, I was shocked that fewer than a quarter of the titles submitted were by women. Why? Many, like Sanmao, have excelled in the field, and many, also like her, long for the freedom of travel and feel 'a life plain as porridge would never be an option for me'. Martha Gellhorn summed it up when she wrote, as I cited in the Introduction, 'The open road was my first, oldest and strongest love.' I don't think the business of having babies and tending the hearth has much to do with the scarcity of women travel writers. You can scale back extreme environments and cart the tots along, or take a break to write something else – biographies of travellers, say. A change of gear can stoke creativity. Men have always seen the travel book as a means of fostering Tarzanian appeal – though any kind of book will do. Hemingway called his 1950 novel *Across the River and into the Trees*. I prefer E. B. White's version, 'Across the Street and into the Grill'. I also said in the Introduction that the 'I've-Got-A-Big-One' school has ceded ground. But television still churns out programmes depicting random blokes with beards yomping across the jungle, seeing how dead they can get towing sledges across polar wastes, or sewing up their own thoracic cavities with a boot-making kit (I reviewed a book by the man who actually did that in Sumatra). It's not, guys, about beating a landscape into submission like a mammoth outside the cave. Dea Birkett, author of *Serpent in Paradise* (1997), a travelogue I admire about Pitcairn, has said:

Travel and adventure have become confused. To get a travel book published, you don't have to see something differently, you have to do something differently. A man may paddle up the world's longest rapid without an oar, with a gutless friend in the back as a foil, and you might have a bestseller. Women tend not to have these sorts of adventures, because they consider them pointless. In a market that values stunts over

looking and revealing, women travel writers will always be at
a disadvantage.

Like television, the big screen has hardly excelled itself.
Beyond *Thelma and Louise* (released in 1991), one can think of few
female travel films like *Wild* (2014) in which Reese Witherspoon
plays Cheryl Strayed, author of a book about a redemptive hike
(she invented the surname, because *straying* was her business).
Redemption seems to be all right as long as a bloke at the start
creates a need for salvation (*Eat, Pray, Love*). Producers used to
say, when I quizzed them, that there was market resistance to
women having adventures that aren't domestic. My late agent,
who represented me for twenty-three years, nurtured most of a
brilliant generation of travel writers who rose to prominence in
the 1980s, including Gellhorn and Paul Theroux. His business
partner asks, 'Why is it that of all genres, travel writing has
remained so predominantly male – even macho? Are we really
only interested in seeing the glory and horror and complexity of
the modern world through men's eyes?' The twentieth century did
produce role models, as I have indicated. Sybille Bedford found,
in her book *Don Otavio*, the story of a Mexican journey which I
mentioned in the previous chapter, a way of anchoring her ideas
in landscape – what all the best writers do. In an unswept room
in Pátzcuaro, the one with a rusty tap that dripped, and someone else's hairpins on the chest, everything damp, the author is
reduced to 'drinking tequila in speechless gloom' and dining off
food that 'tasted of swamps'. When it happens to me, I think of
Pátzcuaro, and resolve not to be miserable, but to make something of it, like Bedford did.

Paul Theroux partly inspired the seventies renaissance when
he set off by train to India, and then to Patagonia. In the aftermath a superfluity of travelogues took a bogus motif as their
central theme. You know the kind of thing: *Up Everest with One
Hand Tied Behind My Back*. Dea Birkett hinted at this earlier. The

trope reached its logical conclusion with Tony Hawks' bestselling *Round Ireland with a Fridge*, for which the author hitchhiked the length of the country with an item of white goods. As for Theroux – at his best he is still (among) the best, and of the other men of his generation who made it in the field, the comic energy of his prose sets him apart.

I often hear it said that tourism has murdered travel writing. I don't think so. Mass travel has liberated the form. No amount of package tours will stop the ordinary quietly going on everywhere on Earth. I remember a young Zhuang woman in Ping'an watching the Olympic showjumping on a snowy television while brushing her false topknot like a horse's tail. In Chile I found my weekly trawl round the supermarket exquisitely gripping: watching women decide between this jar of *dulce de leche* or that one, weighing out their chirimoyas, loading up with boxes of washing powder. In Greece a decade earlier I had joined girlfriends at their weekly weigh-in at the local pharmacy (domestic scales had to wait for more prosperous times). The sight of people paying bills or buying a hand of bananas – those are the most emotional parts of travel. You glimpse a life in all its minute detail. I was always looking for the poetic in the everyday, caught from a fleeting but revealing angle. I once saw a man outside Sainsbury's in Camden Town picking up a carton of milk from a discarded crate, popping it open, and sniffing it. You don't *actually* have to be off the map. Obsessive curiosity is a requirement of the travel writer – and of the biographer, novelist and poet. In *Paris: a Poem*, the story of a *flâneuse*, Hope Mirrlees notices 'hatless women in black shawls carrying long loaves' and 'little boys in black overalls' with sticky fingers. The significance of the trivial is what makes a book human. Of course, the writer must select the fertile fact; Craig Brown recently noted certain biographers' ability to pack in the untelling detail ('her primary school teacher had recently moved into a bungalow'). Out there on the road, I have often found that the most aimless and boring interludes yield, in the long run, the most fecund material. Every journey creates energy, joy and, above all, hope at some point. There is always a dash of human dignity to lift a story out of absurdity and farce, however ugly the background. The world everywhere and simultaneously is a beautiful and horrible place. The notion that all the journeys have been made is just another variation of the theme that the past exists in Technicolor while the present has faded to grey (perhaps Payne's grey?) – that everything then was good, and everything now is bad. A theme, in other words, as old as literature. I add the point that there are no package tours to the Democratic Republic of Congo, still the heart of darkness, or to the parts of Saudi Arabia where women live in a perpetual ethical midnight.

I cite Victorian 'lady' travellers with caution. They have evolved into a cipher for ridicule as they crest unnamed pamirs, parasols aloft, or sail through malarial swamps trailed by a retinue of factotums; in my mind's eye I see a Monty Python special in which they semaphore despatches from a windy mountaintop ('Inadvertently sawed legs off, but chin up!'). They have become the humorous flank of a business men did and do properly. Except the caricature is (often) not true. Kate Marsden, a writer who has not yet had her due, beats even Papa in the title department. In 1893 she published *On Sledge and Horseback to Outcast Siberian Lepers.* I recognised something of myself in Isabella Bird, whom I mentioned earlier. She yielded to the impulse to flight, escaping both demons and the curtain department; she craved the anonymity of the open road; she was in many ways absurd. Between 1894 and 1897 Bird travelled for fifteen months in China, eight of them on the Yangtze and its tributaries and in the regions watered by them. Standing two inches short of five feet, Bird had a fragile build and a fine, beaky nose, a study in nominative determinism. Her father was a man of the cloth, a fanatical dispatcher of missionaries to heathen lands and a Sabbatarian, which meant he campaigned against Sunday trading. The reverend homeschooled his two daughters. Isabella, born in Yorkshire

in 1831, was a sickly child who grew into a sickly adult. When
she was twenty-three, her father gave her £100 to visit relatives
in America, believing that a change of climate would restore her
health (either that, or wanting to get the stench of sickness out of
the rectory). She was off. A pattern established itself: ill at home,
Bird flew the nest and recovered, only to collapse again when
soft furnishings loomed. Some inner fire burned, one that never
found the right fuel in a conventional setting. In the 1860s the
Birds migrated to Scotland, and Isabella and her sister Henrietta
flitted between Edinburgh and Tobermory on the Isle of Mull in
the Inner Hebrides. Bird was almost always ill during this period,
suffering fevers, lesions, rashes, nausea, tumours, chest pains,
muscle spasm, hair loss, neuralgia, jaw ache, insomnia and con-
stipation, and she was often too frail to hold her head up without
a steel support. According to a modern specialist, she probably
had carbunculosis, a deep staphylococcal skin infection, in her
case on the spine. In a fruitless attempt to cure a generalised
infection, she had all her teeth out. As a proper neurasthenic of
the period, Bird drank alcohol for health purposes. 'I have wine
at eleven,' she reported, 'lunch and beer at one, wine at four taken
out if we are driving, dinner and beer at six, tea wine and bed at
nine.' Bring it on! But booze failed to solve any problems (does
it ever?) and in 1872 despairing doctors recommended a world
tour. She was forty-one.

So Bird became a professional traveller. When she was alone,
she was safe in the travel writer's natural habitat; nobody would
ever recognise her. Anonymity can be a powerful aphrodisiac.
It was to Bird. She was a transgressor, an outsider everywhere,
a cuckoo. Her book *A Lady's Life in the Rocky Mountains* (1879)
recounts a trip to Colorado largely on horseback in which she
fell in love with 'Rocky Mountain Jim', a grizzled figure with one
eye, blue as a gas jet. Jim Nugent was a scout, a trapper and a
fisherman who galloped into Boulder with three hundred pounds
of trout in his panniers. Why did she let him go? Why didn't she

just stay, as so many have? Instead Bird became a fixture on the publishing list of John Murray, then a family-owned London firm specialising in eccentrics. Both John Murray III and John Murray IV were friends of Bird's for decades, and the former taught her to ride a tricycle outside the firm's office in Albemarle Street. Bird achieved fame, even dispensing advice on foreign lands to prime minister Gladstone. When the home front induced illness again, Murray IV dispatched Bird on another world tour, in the course of which she stopped in New York and San Francisco on the way to Asia, Egypt and the Holy Land. She was in fine health until the ship back to Britain. Within a year of her return, her sister Henrietta died of typhoid. Bird was distraught, and coped with her grief by marrying her sister's doctor. She was fifty, he was forty, and they had the wedding invitations printed on mourning stationery. Bird even wore a black wedding dress, seeking immolation on the marital bed. Her health again careered downhill, though she perked up when the husband expired of pernicious anaemia.

She had, she said, no intention of writing a book about China. She had published eleven volumes already. Instead, for the first time she planned to create a photographic record. And she did. The pictures range from the intimate (a woman begging from the entrance to a hut) to the epic (a sugarloaf mountain in Siao Ho) and include many skilfully framed compositions, my favourite the bridge at Wan Hsien (now Wanzhou), arching across the valley floor like a swan's neck. Bird developed and printed her photographs in the boats on which she travelled up the Yangtze. It is hard to believe what she achieved under those primitive conditions. 'Night supplied me with a dark room,' she writes in the book that in 1899 emerged from the journeys after all. 'The majestic *Yangtze* was "laid on" [she means it had running water]; a box served for a table: all else can be dispensed with.' She lined a tiny cabin with muslin curtains and newspapers, and, as light from the opium lamps filtered through the chinks, tacked up blankets and slept in her fur coat. She leaned over the gunwale and held the negatives in the wash to fix them. When not on the river, Bird travelled in a sedan chair carried by three bearers and stayed in inns or houses, chiefly in Sichuan. She had only one bad night, when nine sows were living on the floor below her, perhaps ancestors of the ones I sat on to eat breakfast. 'The groaning,' she wrote, 'grunting and rooting were incessant' – both noises and smells floated up through the floorboards. When her European clothing fell apart, she dressed in Chinese garments and straw shoes, carrying her luggage – five porters carried it, actually – in two deep, square bamboo baskets. On one occasion she came upon a young married woman 'lying apparently dead' having tried to commit suicide with opium as a result of a quarrel with her mother-in-law. 'The villagers appealed to me for remedies,' she wrote, 'which I succeeded in forcing down her throat, and put plasters of hot vinegar and cayenne pepper behind her ears ... I had a bad quarter of an hour before she became conscious, for, had she died, the opium would have been acquitted, and the blame would have been laid on the foreigner.'

It was not one long idyll – one wonders if any trip ever is. 'The crowd caught sight of my open chair,' she writes one evening, 'which, being a novelty, was an abomination, and fully two thousand men rushed down one shingle bank and up the other, brandishing sticks and porters' poles, yelling, hooting, crying "foreign devil" and "child eater".' They proceeded to stone her. *Stoning?* My only problems in China were internal: people generally took no notice of me whatsoever (no Mandarin *moni sous* there), but my vast size, grotesque appearance and pallid skin made me look and feel freakish compared with Chinese women (and men). Bird's enthusiasm never wavered. She can't have felt invincible. People think you do, but you don't. Despite another stoning, the Chinese made a favourable impression. The population of the Yangtze valley, which she put at between 170 and 180 million, was, she wrote, 'one of the most peaceable and industrious on earth', able custodians of 'an elaborate and antique civilisation'. These people were not the 'Yellow Peril' but the 'Yellow Hope'. Throughout her pages, Bird writes of foreign nations competing for influence in China. In 1898, the year before *The Yangtze Valley* appeared, Britain claimed the basin as its 'sphere of influence'. Bird ends her book, 'China is certainly at the dawn of a new era. Whether the twentieth century shall place her where she ought to be, in the van of Oriental nations, or whether it shall witness her disintegration and decay, depends very largely on the statesmanship and influence of Great Britain.' *Dulce est desipere in loco.*

As for the new dawn I mentioned earlier – dusk had fallen in the gardens of the West. Capitalism had already soured when the Big Bang deregulated London's financial markets in October 1986. The cream of hope had soured into the curds of disillusion. The Thatcher government had transformed the institutional culture of the money business, leading to multi-zero bonuses, sundry other follies and, ultimately, the 2008 crash. Firms like the Wiltshire Friendly Society from which Nana and Pop had drawn the half-crown Christmas gift for me no longer existed, and nor did their 'collaborative capitalism' – the word 'Friendly' was obsolete in the new financial universe. It was all yachts now, and a vertiginous rise in pay differential. Whether the current appetite for responsible and ethical investment, and for corporate governance, will play out to any significant degree over the next years I don't know. But I doubt it. Meanwhile the 2008 crash also stoked populism, both right and left, as did the Eurozone crisis which followed it, and the migration issues of 2015. David Cameron had been at Number 10 for two years when I followed China's Qinghai–Gansu–Inner Mongolia arc, and unlike Gladstone with Bird, he had not called on me for advice – unwisely, I think. Two decades had passed since that bright morning in a Shanghai hotel. The phrase Blair's Babes, rotten at the outset (as a phrase, not a concept), was long in the dustbin of history – the landfill one rather than the recycling bin. I came of age in the Cold War. Was I growing old in Cold War II? Most western scholars in the China game dislike the analogy, but increasingly, I saw almost everyone accept that China under Xi Jinping, to quote Niall Ferguson, 'is behaving in ways that recall Churchill's characterisation of the Soviet Union's "expansive and proselytizing tendencies" under Stalin'. Warnings that China has a grand strategy for world domination, Ferguson also notes, 'are now a favourite with publishers', and I had seen the evidence on the serious books' pages. Left and right for once agreed, to cite one of the books, that 'Beijing's ultimate objective is to displace the US order globally . . . to emerge as the world's dominant state by 2049'. Both sides agreed too that Deng Xiaoping's maxim of *Tao Guang Yang Hui* ('hide one's capabilities and bide one's time') has run its course, and that hiding and biding are over in the age of Biden. In every country I visited in the three years before Covid, I came across evidence of Xi's Belt and Road Initiative, projecting Chinese resources, soft power and propaganda beyond the Indo-Pacific. China was 'moving closer to the world's centre stage', as Xi had put it as far back as 2017, which is by now the Cambrian, or at least the Ordovician, for China-watchers. Domestic inequality flourished alongside this expansionism and China's internal weaknesses continue to grow – Beijing can't hide the figures, just as it can't hide the Uighurs. A Chinese invasion of Taiwan will determine what comes next.[1]

---

1. Coverage of the 2022 war in Ukraine often invoked Cold War II. So perhaps I was to grow old in the cold after all.

8

Don't Wake Sofia

Russia

The power point in the carriage corridor was under the (now electric) samovar. A cable snaked out of it, beneath the carpet and into the *kupe* adjacent to my own. There two bullet-headed men sat comfortably, selling access to a multi-socket bar they had brought on board for the purpose. Private enterprise had captured the phone-charging market on the Trans-Siberian.

The technical feat of this railway reveals itself in winter. Russian engineers might as well have laid tracks across Antarctica. A monument in Vladivostok states that the line covers 9,288 kilometres, but the route has since been tunnelled, shortening it a little. Trans-Sib 80 calls at Ekaterinburg on its route from Moscow, and I boarded there one December evening to ride the rails to Irkutsk, a journey that takes three nights and two days, brushing the top of Kazakhstan before dipping towards Mongolia. While I waited on the platform for the *podranitsa* to tick my name on her clipboard, my nostrils froze. The station echoed with the sound of steel scraping steel. A mural depicted Gary Powers parachuting to earth accompanied by a shard of U-2 fuselage painted with a Star-Spangled Banner.

I had spent several days in Ekaterinburg. The gigantic thermometer at Koltsovo had revealed a tropical minus three when

I landed from Moscow. Outside my homestay toddlers ran about in ski suits, and people brushed windscreens clear of ice. Ekaterinburg has only one skyscraper, and nobody likes it: they call it 'the beer can'. Yeltsin was born in Ekaterinburg, and rose to become regional party boss. Like the beer can, he is unpopular. When Putin insisted there should be a Boris Yeltsin Street in the city, Ekaterinburgers protested, as nobody wanted to have the former president in their address. So compromise was reached, and Yeltsin presides over a commercial street where nobody lives.

The stewardess on the train, at first as icy as the landscape beyond the window, softened when she heard my faltering Russian and noted that I was alone in the *kupe* in the company of a couple of vodka-glugging engineers. This pair spent most of the journey in the smoking area between carriages, an unheated zone which achieved the feat of being both kippered and crusty with ice. During the first night we woke to a thunderous knocking beneath our heads. Had the train smashed into shards of another downed spy plane? In fact we were in a station, and railwaymen were using poles to dislodge ice on the wheel hubs. When I opened my eyes again it was light and people were scurrying round a large station yard, furred head to boot. The Trans-Sib traverses seven time zones, one of which jumps two hours; in winter the sun rises over one end of the line as it sets over the other. Railway clocks, however, remain permanently on Moscow time, as does the timetable. My iPhone remained loyal to the local hour (until it ran out of power and for a single afternoon I took an anti-extortion stance), but it could not tell me if time had changed while I slept. So I entered the unmoored zone. We were crossing the taiga, the boundless coniferous forest forever haunted by the image of a man in chains. A dilapidated cottage and outdoor privy occasionally emerged from the larches in a mist of falling snow. As soon as Tsarevich Nicholas laid the first stone for the Trans-Siberian on 31 May 1891, problems queued up for recognition. Paucity of labour, non-existence of supplies – despite thousands of miles of taiga, the wood was too soft to make sleepers – then there were the mighty rivers; weather; and permafrost. The railway nonetheless represented the final colonisation of Asian Russia by the imperial east, and construction workers, many of them prisoners, transformed Siberia into a magnet for immigrants. Five million Russians disembarked in the two decades before the First World War. You wonder how such dramatic economic development has been so effectively squandered. According to Michael Stuermer's *Putin and the Rise of Russia*, well into the new millennium 40 per cent of homes in the countryside and small towns were without running water, sewers, telephones and often electricity.

Meanwhile in overheated *vagon* nine, *kupe* six, Aliek and Pavel were en route to Novosibirsk to purchase machinery for their voltage-tower-making plant. Together we watched the loamy plains of the Urals congeal into Siberian forest. Ropes of freight trains passed in the opposite direction, heading west from the coal basins to the smelting plants of the Urals. The train followed the day from village to village – first lights, shovelling snow, vans beetling at noon, children walking home from some lonely school, lights snapping on in disintegrating *izbi* – then more larches and beech trees. Chekhov said that only migratory birds know where the taiga ends. The Trans-Sib stops for half an hour when it does stop, enough time to bundle up and take a stride, and at Omsk the *podranitsa* alighted to purchase snacks from itinerant platform vendors. These she sold on to us over the course of the remaining journey. Aliek bought a table-tennis bat of dried fish from a patrolling *baboushka*, wiping his hands on the net curtains of the compartment after he had eaten it. A tray of hot doughnuts came on, infusing the carriage with the aroma of spicy sugar, and then it was east to the Barabinsk Steppe where Kirgiz nomads once grazed herds from the Irytsh to the Ob. At Novosibirsk a sculpted Christmas ice village glinted in winter sunlight on the frozen Ob. Once a simple stopping-place on the brutal *trakt* road, Novosibirsk had swollen into the capital of Siberia and the third biggest Russian city, its train station the largest in all Russia. On our platform, alongside Number 80's red engine, boxes of electronic gadgets changed hands, bricks of cash stowed quickly in a plastic bag. The engineers had left the train, off to buy aluminium conductors, and Galina replaced them in the carriage. A clear-skinned beauty of about thirty, she was heading home to Krasnoyarsk from a forestry conference; a selfie she showed me depicted her threatening a tree trunk with a jumbo hypodermic. Later, as she ate a Pot Noodle hydrated from the samovar, she took my notebook and wrote the Latin names of the trees we could see beyond the window.

At Cheremkhovo, the site of one of the biggest coal deposits in Russia, even the snow on the platform was black. Russian trains at that time hauled well over a billion tons of freight a year, a quarter of it coal and a fifth oil and petroleum products. Since then, coal freight has increased, and as I write, war in Ukraine has put further pressure on the supply chain. Five years after my Trans-Sib trip, at the COP26 coal plenaries, Russian delegates (Putin did not attend) pushed back behind the scenes. The country had recently voted against coal-reduction pledges at the G20. Everyone knew that Cheremkhovo and the other production centres in the Irkutsk basin would be trucking out black gold long after the placards of the Glasgow protesters had sunk into landfill.

My ticket included two meals a day, and as a grey Siberian morning brightened, I ate the last: a few spoons of instant mashed potato served on an indented plastic tray with a finger of white fish and thirteen peas. On the platform, a platoon of young Russian soldiers, green plastic sandals swinging from packs, said goodbye to mothers and girlfriends. Once the train pulled out each woman stood like Gurov on the Yalta platform as he 'gazed into the dark distance' after the fair-haired lady and the Pomeranian have left for Petersburg in Chekhov's story 'The Lady with the Little Dog'.[1] The *podranitsa* yelled that facilities were temporarily closing, and set about her cleaning ritual with gusto. She unhooked the rail from a net curtain in the corridor and used it to plunge the lavatory, though she did first slip off the net.

The trip was among many in Russia in my mid-fifties researching a book in which Golden Age writers guided my travels – Pushkin to Tolstoy, via Leskov. They perceived their work, and that of their peers, as a moral tool, one that helps address the question, 'How are we supposed to live?' It was a question worth asking, even as one sailed into the sixth decade. Perhaps especially then. In addition, life continued its pursuit of art. Donald Trump entered the White House when I was researching the Russia book. I had just read Dostoevsky's *The Idiot*, so it was impossible to feel surprised. Besides, Russia was the first country I had ever visited. I indicated earlier that my childhood was characterised by homogeneity, but in fact one thing distinguished us from other families. My mother believed she had been Russian in a previous existence. The conviction was based on a viewing of *Doctor Zhivago* at the Henleaze Orpheus, but mother was sufficiently keen on discovering the Soviet within that she signed up for evening classes in Russian, and often took me along. I remember us both sitting in baffled stupor on the minuscule chairs of a local infant school while a bouffant-haired teacher scrawled hieroglyphics – or was it Cyrillic? – on the blackboard.

One day, my father, who was and is a builder, saw an advertisement in the *Daily Express* for package holidays to Moscow. The concept of the package holiday was new; we had never been

---

1. The story's title is often translated as 'The Lady with the Dog'. Russian has a word for dog, and another, a diminutive with one extra letter, meaning little dog or doggy. Chekhov chose to use the latter, and to my mind it conjures a wholly different image and a more appropriate one – the reader sees a creature more accessory than hound, a picture in keeping with the story's tone of delicacy and restraint. The tale is sometimes also translated as 'The Lady with the Lapdog'. In 1996 the London-born short-story writer Clive Sinclair brought it up to date as 'The Lady with the Laptop'.

abroad; Russia was as remote as the South Pole. But before I knew it the money had been found, and in September 1971 we made the long journey across to Heathrow, crawling along A-roads between Swindon and Maidenhead, where the M4 did not yet exist. I was ten.

It was late at night and explosively cold when we were herded from the airport bus into the grand entrance of the Rossiya Hotel, a twelve-floor, four-thousand-bed block of Soviet concrete in the historic Zaryadye district, hard by the Moskva River. In this monstrous edifice, which had its own concert hall and which later entered the *Guinness Book of Records* as the largest hotel in the world, the watchful authorities billeted westerners. I remember the synthetic cherry smell of floor wax. Premier Kosygin had nominally ruled the country for seven years, but First Secretary Brezhnev was in charge. Although the Soviet Union was not the economic basket case of later years, Brezhnev's reforms were failing, and stagnation loomed. To us – hardly accustomed to luxury – everything in Russia looked grey, tatty and worn-out.

We awoke to a vista of Red Square, our grimy window facing the monumental bronze of the patriots Pozharsky and Minin, greenish with age but still plotting the expulsion of the Poles. Having piloted us safely to the cavernous dining room, my father engaged a waitress in the first of many high-decibel conversations in pigeon English. He conveyed the message that we wanted eggs (can he really have clucked and flapped his arms?), waggling three fingers to indicate that we all wished to have an egg. Half an hour later, just as nine fried eggs arrived at our table, twenty-odd men in grey suits with cardboard shamrock buttonholes trooped into the room and sat down. My parents recognised them immediately as the Northern Ireland football squad, in Moscow to play a European Championship qualifier. It was as if the gods had descended. Derek Dougan! Dave Clements! Tommy Cassidy! And of course, the Promethean figure of George Best himself, sheepishly tackling a plate of rubberised fried eggs just like us. All talk of Romanov ancestry vanished as groups of players stopped at our table for a chat. (Foreign tourists were a rarity in Russia then.) Best was twenty-four, and already established as one of the most dazzling wingers the beautiful game had ever seen. He had been wearing the Number 7 shirt for Manchester United since making his debut as a shy seventeen year old with a Beatles haircut. The season before the Moscow fixture he had netted his now legendary six goals in a cup tie against Northampton Town. Drink had not yet taken hold, and Best's devilish good looks, sky-blue eyes and soft Belfast vowels had won the nation's heart. In Moscow he let others do the talking. When player-manager Terry Neill asked if we'd like tickets for the match at the Lenin Stadium that night, Dad almost fainted clean away. Later that morning, in the hotel foyer, we posed with the stars in front of a small mob of British newspapermen in belted mackintoshes (they had not yet evolved into paparazzi). Best was noticeably less forthcoming with the press than some of his teammates. He certainly didn't care for being photographed. But it was his face they wanted. 'Put your arm round her, George!' a snapper shouted when he spotted me close by. 'She's a bit young, even for me!' he called back. I can hear him saying it now.

Wearing all the clothes we had packed, that night we sat alongside tens of thousands of Muscovites under the malfunctioning floodlights of the Lenin. Everyone smiled at us. Dad wanted to buy a programme as a souvenir, and attempted to convey the fact to a bear-like man in the adjacent seat. Once the ruble dropped, the man leaned over into the next row, snatched a programme from an innocent spectator's hands, and gave it to Dad. The greens went down 1–0 after a penalty decision went against Terry Neill. The next day, our tribe back in Bristol opened the *Daily Express* to see us grinning goofily alongside Georgie Best. We returned to a heroes' welcome, and still talk of the event today.

Since that trip I had visited Russia on several assignments, including two long ones for the book about the circumpolar

Arctic. Now my children were older I had leeway to make longer journeys, even in term time, and of course I could still cart them round with me in the holidays. The Russian literature of the nineteenth century stands alone in the history of letters, and George Best would have understood that I needed now to pay homage to another team of masters. I began to learn Russian myself with the aim of getting through a Chekhov short story ('a scale model of the world') in the original. Chekhov's prose is clear and his short stories short, so it was not an unreachable goal, and there aren't many words in 'The Lady with the Little Dog', the one I picked. The reader will have noted my fondness for trains. I decided to make the railway my preferred mode of transport for the embryonic book. Aeroplanes don't represent travel in any meaningful sense – you get on at A and start over at B, doors closed to manual even for a writer operating in a genre which Jonathan Raban, cited in the previous chapter, refers to as a 'notoriously raffish open house'. In a car or bus you can't stand up, which constricts the mind as well as the body. The railway carriage, on the other hand, is a sealed world from which one agreeably observes that pesky other world whizzing or crawling past the windows before obligingly vanishing (I called it an 'unmoored zone' earlier). Since the days of *Hindu Choo Choo*, the project languishing in the sidings since 1997, I felt I had not travelled enough on trains, at least in foreign lands. In the UK I took to the rails all too often as in 2015 I gave up my car as a (feeble) gesture towards greenery. Anyway, in Russia I hopped into a carriage whenever I could, and, as with many experiences in life (one thinks of love affairs, or car ownership), the first journey was the most memorable – on the Moscow–Petrozavodsk express, a nine-hour ride to the heart of Karelia, Russia's swampy northwestern republic. It was the time of white nights, the sky pearly till dawn. At Shlisselburg a bridge carried the 658AA over the low-water Neva. At Svir, I see from my notebook, I purchased a bag of Karelian dried smelt from our *podranitsa*, and their fishiness perfumed the *kupe* all the way to Petrozavodsk.

I liked the variety of Russian trains – their constellated numbering combinations, for example, and their byzantine ticket options. I can imagine myself a Russian spotter, swaddled in sheepskin alone at the end of a platform in Tayga with a red-and-black notebook in my gloved hand. Rail differentiation is a function of a country's size, but uniquely, Russian trains showcase the ways in which entrepreneurial flair fills holes in a broken system. I have mentioned the phone-charging private enterprise, the deployment of a curtain pole in the absence of a lavatory brush, the contraband changing hands on snowy platforms, and the low-level profiteering of staff selling on packets of horrible sweets. In addition, trains have embedded themselves in twentieth-century Russian history. Nicholas II signed his abdication papers in a panelled compartment of the Imperial train at Pskov. But the country came late to railways. The first line opened in 1837, between Moscow and Petersburg, and the first station was called a *vokzal* after the English 'Vauxhall', and it included a concert pavilion, like the one in Vauxhall Gardens.

(The word still means 'station' in Russian.) By the early 1880s, Russia had laid 23,335 kilometres of track – roughly the same amount as Great Britain, a country seventy times smaller; it had not yet joined the country coast-to-coast, whereas the US and Canada both had transcontinental rail links. But Russia caught up. By 1904, Muscovites could travel to the Pacific on a train. During the revolution, Bolshevik representatives of the People's Commissariats commandeered trains to take propaganda to the villages. Many people had never seen a film before, and they paid for their tickets with eggs. Agit-trains carried a printing press to allow customised posters to be produced and hurled out of the windows. Construction standards were at best suspect. In 1882, all Russia recoiled at the news of a crash at Kukuevka on the Moscow–Kursk line: an embankment collapsed, burying hundreds alive. Writers began to express misgivings about the role of the railway in the rush to join Russia together. Dostoevsky did in *The Idiot*; the smoke in Turgenev's eponymous novel is railway-engine smoke; the first phrase of Chekhov's *The Cherry Orchard*, 'The train has arrived,' adumbrates unsettling change.

After 1959, electrification crept in all directions, though the 658AA line on which I took my first journey was not wired up till 2005. Sometimes wheels turn slowly. The few high-speed trains on today's network are no different from TGVs or *Shinkansen*. The German-built *Sapsan* (Falcon), for example, transformed travel between Moscow and Petersburg when it opened in 2009, and remains the only profitable passenger service in the country. The *Sapsan* runs fourteen times a day. I once jumped on one at Tver, a stop on the Volga, and plugged in my laptop at the table along with everyone else – heads down all the way with no samovar in sight. Elsewhere, as one travels further from the cities the view from the window reveals the economic and social polarisation within Russia. One spring I boarded a Number 22 sleeper at Petersburg's Ladoga for a sixteen-hour voyage through the birch forest in a first-class *spalny vagon*. Overnight, birches grew shorter (as did people) and the air colder. Cars also shrank as the train raced north: rust-bucket Ladas instead of the BMWs of metropolitan mobsters. Inhabited houses were virtually derelict. The landscape had just hung on through winter and was browned out and exhausted. No factories stared through eyeless windows, and no toppled statues of Lenin sneered skywards in Ozymandian reproach. Just the noble rot of the backwoods and a woman in a housecoat milking a goat. A sway at a curve; gabled snowmobile lock-ups on the shore of Lake Vyg; and, at last, the White Sea port of Kem, halfway between the Kandalaksha Gulf and Onega Bay.

On that trip to Kem, the Number 22 stopped at Segezha, a Karelian barracks-style prison Russians call 'The Zone'. Mikhail Borisovich Khodorkovsky served the last part of his sentence in Segezha, making paper folders for $15 a month. Family visits to Russian prisons are infrequent, but when they happen they last three days, as it takes everyone so long to get anywhere. Across the bleak platform, a hunched couple hauled luggage in relay.

Russian Railways, a megalithic monopoly, transports more than a billion passengers a year, most of them suburban commuters. But swathes of the country remain without rail access. Far eastern Chukotka covers three-quarters of a million square kilometres, with no soil in which anything can grow (as one traveller said of the mountain basins of northeastern Siberia, 'twelve months of winter, and the rest is summer'). This is the region Sarah Palin saw, she said, during a vice-presidential campaign which rendered satire obsolete, from her bathroom window. Not only does Chukotka have no rails, beyond the capital, Anadyr, it has no roads either. When I was there (the region is closed to foreigners, but that is another story), a resident had taped a note to an Anadyr lamp post offering his flat in exchange for a single plane ticket to Moscow.

One July in the capital I settled into a *kupe* on the night train to Pskov, the handsome city close to the Estonian border where the tsar signed out. Two young women were already exchanging vertiginously high heels for railway-issue slippers, and when they had finished, they set out elaborate picnics on a lower bunk. I had learned to bring contributions to what inevitably turned into a communal meal, especially when travelling in third-class *platskartny*, the open-plan dormitory cars resembling an amalgam of boarding school and refugee camp. People say Russians are unfriendly. It's true that they rarely smile, and can maintain a stony mien for an unfeasibly long period. But once the shell breaks, Russians are very often warm, hospitable, open and even funny. ('You must trust and believe in people or life becomes impossible,' according to Chekhov.) Once I took an assignment on a boat proceeding southeast from Petersburg across lakes and along rivers before meeting the Volga–Baltic waterway at Moscow. It was a river cruise, and Russian lecturers on national history were not afraid to introduce personal comments. Raisa Gorbachev, we heard, was the first Russian First Lady who weighed less than her husband. I enjoyed the twelve-day trip. I had boarded in the morning at a pier next to Sverdlov Bridge. It was a hot summer, and spam-coloured Russian bodies lay on the grassy banks outside the Peter and Paul Fortress. *Truvor* was not raising anchor till the evening, so I got off again and took the metro to the courtyard where Raskolnikov killed the old money lender with an axe in *Crime and Punishment*. Dostoevsky, above all other writers, captures the other side of Petersburg – dirty bars, crummy lodgings, misery. It's a long way from the divine baroque created by Catherine the Great's architect Francesco Rastrelli, where a hot Italian palette replaces the greys and blacks of Dostoevsky country.

Meanwhile, back on the Pskov train the two young women and I ate *kolbasa* sausage with the ubiquitous cucumber (ever had a meal in Russia without cucumber?) as the butterscotch light of a late midsummer evening settled over Zelenograd, a Russian Silicon Valley, and after that over forest, the odd factory and sharp-roofed chalets set in fields of flowers. Picnics notwithstanding, some long-distance train tickets include daily meals, as indicated by the thirteen peas on the Trans-Siberian. In the summer of 2012, on the 644 from the Black Sea inland to Pyatigorsk and the old spa towns of the Caucasus, I had my children with me. The menu announced chicken and pasta. Reg's portion consisted of a hen's foot. That line runs along the Black Sea Riviera, and as the track was not laid over the mountains, one travels east into Caucasian heartlands via two sides of a triangle. Just an hour out of the horror story that is Sochi, the track pierces undeveloped land, crossing inlets alongside beaches where people were still bathing at nine at night. 'Everything is beautiful in this world when one reflects,' Chekhov writes in 'The Lady with the Little Dog', 'everything except what we think or do ourselves when we forget our human dignity and the higher aims of our existence.' When we woke, we were travelling through arable land quilted in solid greens and yellows and dotted with baled hay. Trees grew through the roof of abandoned cement factories. The 644 yielded my only male of the *provodnitsa* species – the *provodnik*, a figure similar in all respects bar hairstyle. On the return trip from Pyatigorsk to Adler a week later, the same steward, standing on the platform, flung his arms wide and greeted us like family.

On the earlier trip I had left the Trans-Siberian in Irkutsk, the city known in the nineteenth century as 'the Paris of Siberia', a moniker which has outlived its usefulness. The main square displayed Christmas ice sculptures like those I had seen on the frozen river Ob. Due to unseasonal weather, some had begun to melt, unfestively shedding arms and noses. For centuries Irkutsk was a place of banishment. From a bell tower of the Church of the Saviour near my lodgings, looking out through iced lashes to the white glare of the forest beyond, I saw how far away Moscow must have seemed to an exile. When I set off by car the next day down the Baikal Highway, it was minus eight. While temperature fluctuation is a characteristic of a continental climate, Sergey, my driver, said that on the same day last year the mercury fell to minus thirty-five. (He remembered, as it was his wife's birthday.) At any rate he left me in the village of Listvyanka, where I was to spend two days walking along the shores of Lake Baikal. Waves broke on the pebble beach (Baikal does not freeze till the end of February), the snowy Hamar Daban mountains stood out sharply between blue sky and silver lake, and the whiff of fish drifted from the market. A woman hauled a bucket up from a public well. The Listvyanka streets had a pleasant resort-out-of-season feel; even the caged bear on the main street was hibernating. Lake Baikal itself is an agglomeration of superlatives. It holds more water than all five American Great Lakes combined; is the deepest lake in the world (on *average* 744 metres); is larger than Belgium. Twenty-five million years old, it nourishes more than 3,500 species of plants and animals, 2,600 of which can be found nowhere else. Limnological nirvana in the shape of a banana. On the second day, snow was swirling across the beach and a sea fog had swallowed Hamar Daban. After a slog into the wind along the shore, I turned off, and in a kind of epiphany, the sun came out, the wind dropped and I passed a man smoking *omul* fish on a brazier. When a dog started howling, all the hounds in the valley set up. By the time I reached the lake again, people were sunbathing with their gloves on. I bought a plastic carton of *omul* caviar at the market for 250 rubles (£5), and ate it with my fingers on the steps of a beach hut.

In the most famous scene in all Russian literature a train crushes Anna Karenina as she flings aside her red handbag and sinks to welcome oblivion. Tolstoy's novel associates the railway with death, not progress. Besides the handbag, throughout *Anna Karenina* trains represent the ugly threat of modernity, adultery, nightmare. The old showman even staged his own death at a *vokzal*. But they took the body back to his ancestral estate. Yasnaya Polyana (the name means Bright Glade) lies

193 kilometres south of Moscow in the Tula region; in Lev Nikolayevich's day it covered four thousand acres and the family 'owned' 350 serfs. Tolstoy always said Yasnaya Polyana was 'an organic part of myself', and in his diaries he describes candlelight flickering on the icons in a corner of his grandmother's bedroom, and a serf orchestra playing Haydn as he, a small boy, walked down the alley of beech trees leading to the main entrance. The house, which smelled of polished wood but not fake cherry, displays objects including the master's passport – a sheet of A4 then, with a description rather than a photograph; a whistle with which Sofia Andreyevna, the sainted Mrs Tolstoy, checked the nightwatchman was awake; and a portrait of her father, a man so grand that he sent his shirts to Holland to be laundered. In Tolstoy's bedroom they had preserved his dumbbells. The type of tunic hanging above his bed, worn with a belt, came later to be known in all Russia as a *tolstovka*, and that, together with shirts buttoned with mother-of-pearl, and a pair of high riding boots, gave the room the look of the 2022 Winter Toast catalogue. Tolstoy had a Shavian fondness for gadgets, and a 1908 Thomas Edison recording device in the house played a crackly reel he had made, instructing the children to be good and kind. I actually heard his voice. It was like hearing the cat speak. He learned to ride a bike at Yasnaya Polyana, careering dangerously among the fruit trees, but towards the end of his life he turned away from the world and its bikes to embrace Orthodoxy, despite his troubled relationship with the religion. One wonders what he would have made of the Russian church's supine role today. As I write, Patriarch Kirill has not condemned the Ukrainian war. On 1 March 2022, 176 Orthodox clerics published a letter calling for 'an immediate stop' to the fighting. It was highly unusual, and brave, for so many priests to defy Putin. Kirill did not sign. Instead he invoked, in public prayers, the concept of a Russian 'fatherland' inclusive of the Ukraine.

I had read many times about Tolstoy's death, more dramatic

than any scene in his fiction, but death came to life only when I watched flickering footage on YouTube. An elderly countess steps from a railway carriage onto an empty platform. Assisted by a man and a woman, she is wearing a dark coat with a fur trim and a black hat. The dawning sky is white, but there is no snow on the platform. Puffs of smoke rise above the railway engine. The three walk slowly towards the station house. The countess draws on gloves. As they approach the building, a slim, bearded figure on the threshold sees them, and slams the door. Sofia Andreyevna had been married to Lev Nikolayevich for forty-eight years, and had given birth to thirteen of his children. At four in the morning a few days earlier, on 28 October 1910, Tolstoy, the second most famous man in Russia, had carried a candle into his live-in doctor's bedroom on the first floor at Yasnaya Polyana and woken him, telling the man softly that they had to leave. Tolstoy said, 'Don't wake Sofia.' He was eighty-two, and wished to renounce the world. His wife was of the world. He had left Sofia a note saying he could no longer bear 'the state of luxury in which I have been living'. He wrote, 'My leaving will grieve you. I'm sorry about that, but please understand and believe me, I cannot do otherwise. My position in the house is becoming, has become, impossible.' The two men – the writer and the doctor – rode to the nearest station in a carriage, well wrapped against the cold as winter was shouldering in from the north. They bought third-class tickets to Kozelsk and headed to the Optina Pustyn monastery, thence on to visit Tolstoy's sister in her convent. After that they set off for the Caucasus. But Tolstoy fell ill, and was obliged to get off the train at Astapovo (a station now called Lev Tolstoy). They put him to bed in the stationmaster's house, a mortal body poised to rejoin the minerals.

The countess threw herself into a pond when she read the note, but it was shallow. After changing her clothes she chartered a train to Astapovo, but when she got there her husband's acolytes wouldn't let her in until Lev Nikolayevich had lost consciousness.

It was one of the acolytes who had slammed the door. The footage shows Sofia Andreyevna pressing her face against the window. Tolstoy's followers had been saying for years that she was hysterical. Who wouldn't have been? Lev Nikolayevich was one of the most egotistical men ever to have lived (which is why he was able to give us the impregnable masterpieces *War and Peace* and *Anna Karenina*). Like many great men, he loved humanity but didn't care much for its individual representatives. He had often written about the moment of death and must have known it was his turn. The Master died at 6.05 in the morning on 7 November 1910. News of his illness had reached the press before he expired, and soon hundreds converged on the Astapovo platform, sixty army officers among them. A Pathé man, reporters and stills photographers crowded round the stationmaster's house alongside the villagers. Students across the land demonstrated, expressing sorrow that such an influential critic had fallen silent, and anger at the government he had so openly criticised. Thousands followed the cortège to Yasnaya Polyana and sang hymns while others knelt in the weak sunshine when the bier passed. They buried him under an unmarked grassy mound. When I visited his grave, a chirruping oriole was pulling at a worm on the mound. The bird and I were alone among shafts of midsummer light filtering through a stand of ash. Beyond clumps of bluebells and hollyhocks, yolky lily blooms patterned the surface of the shallow pond where Sofia Andreyevna had hurled herself. A small sign at the edge of the grave ordered, in Cyrillic, SILENT ZONE. The oriole had not read the sign.

In the Russian years I went to many writers' homes, including Mikhailovskoe, where in 1824 Alexander I exiled Pushkin to his mother's ancestral estate on the country's western rim. There I became obsessed with making my guide, Irina, admit Pushkin wasn't perfect. She was touchingly reverent. Pausing in front of a photo of the poet's brother Lev, she said, 'Lev had many gambling debts which Alexander Sergeyevich [Pushkin] paid off.'

'Alexander was a gambler too, wasn't he?'

'This is a chair he must have sat in.'

We hiked over fields to Trigorskoye, a walk Pushkin took every day, visiting friends and dancing in their open-air ballroom. Irina was reluctant to admit that the poet had affairs with the daughters of the Trigorskoye house, and with their mother. But Pushkin was a heroic shagger who had time to write only when indisposed with sexually transmitted diseases. Another poet once wrote to a mutual friend, 'He [Pushkin] is finishing the fourth canto of his poem. Two or three more doses of the clap and it'll be in the bag.'

Homestays, like trains, were a theme of my Russian travels. Elsewhere in the world tour operators control the homestay business, and all too often folk dancers shuffle into view along with the 'traditional meal' and greenwashed guff about 'ecotourism'. Nobody bothers with that in Russia. The business is run by indifferent agencies with fixed prices, offering nothing bar a clean bed. There is no pretence of cultural showcasing. Hosts need the cash, often vacating a spare bedroom and doubling up. Remnants of a centralised economy mean the bathroom mat is the same colour in Anadyr as it is ten times zones away in Petrozavodsk, and the familiarity made me feel at home. I knew where I was when an inadequately plumbed washing machine in a tiny windowless bathroom spooled plastic pipes into the bath under a jerry-rigged nylon line. I enjoyed sprawling on the sofa of an evening with my hosts, most of them hunched over Facebook and moaning about the latest volley of bad news. In the Khrushchev-era blocks of flats, the walls are thin. In one homestay, in Irkutsk, a neighbour chipped in from the other side of the wall to criticise my poor Russian. The earlier flats, built under Stalin, have thicker walls. The thin-walls topic invariably led to a speech I heard from many Russians (some of the 144 million real ones, not the oligarchs): 'Say what you like about Stalin . . . ' Putin had been rehabilitating Stalin for years, extolling his leadership qualities and suppressing knowledge of the murder he ordered – for example – of many thousands of Poles, Latvians, Lithuanians and Estonians under Soviet occupation in 1939–41. Vladimir Vladimirovich had in some ways taken on the role of people's champion. It came as no surprise when Putin bombed maternity hospitals in Ukraine. *What*, he must have been thinking, *would Stalin have done?* But 'people' suspected that their president was a villain, and during the Ukrainian conflict doubts intensified. I believe that the truth about Putin, or an adequate version of it, will filter through to Russians and that he will not endure in the national imagination as much more than he is – a crazed and deluded megalomaniac murderer.

Natasha was a homestay host. We were in her flat in Moscow with the news on, and she said she had heard about Brexit on the same television channel. This new national identity articulated by the British electorate in my name often came up. It had become embarrassing to be British. As a young woman on the road I had promoted my non-American status to curry favour. But the favour had turned into snickering ridicule. Natasha asked if I felt European. I said I did, and repeated the same question to her. 'Yes,' she said. 'My mother's family came from Siberia though, and they were Asian.' She turned the volume down on the set, and with her permission I pressed audio record on my phone. 'Granny lived in western Siberia and was dark and rich and beautiful. Everyone in the village had wanted to marry her, but she chose a poor illiterate redhead with a big heart. My grandparents built up a big farm, with horses and sheep. Then granny died aged thirty, leaving six kids.' When Natasha's mother was seven, a government man arrived at the Siberian farm, gave Grandpa an envelope, and ordered him to ride to town and hand it to the police. At the station, a *politseyskiy* opened the envelope. 'Do you know what this letter says?' he asked Grandpa. 'It says we must shoot you and your six kids.' That policeman was kind, so he gave Grandpa two days. Nobody would take the children (it was too dangerous), so Grandpa put all six on trains to destinations far and wide. Natasha's mother never saw her family again. Grandpa died shortly after, not from Stalin's bullet, but from a broken heart. 'I look for my relatives in crowds,' Natasha said, refilling the teapot with hot water. Her mother had married her father, a military man, in Moscow, and gave birth to Natasha in the Khrushchev era. 'Till 1997,' she said, 'I spent my entire life queuing. I used to go by sledge to the grocery store with my mother early in the morning. We had to wait two hours outside the shop and two hours inside. Mum had to take me, as with a child she could get twenty eggs not twelve, and a kilogram of butter not half a kilogram.' When Natasha was six, her mother towed her to the grocery store on a sledge as usual. When her mother arrived at the store she looked round, and Natasha was not there. The frenzied woman ran back and found Natasha asleep in the snow. Natasha remembered hot tears. One day, a year or so later, a woman queuing in the same store asked Natasha and her friend if she could borrow them, so she would be able to purchase more rations. Of course, the girls didn't fancy yet more queuing. But the woman promised them chocolate as a reward. 'A chocolate! For me!' said Natasha, the lure of the treat still sparkling in her blue eyes. 'The chocolate that woman bought us after we had queued was the tiniest thing, wrapped in yellow paper with a kitten playing on it. We were the happiest girls in the world. Last year that childhood friend came to visit me, and she said, "Natasha, have you ever eaten anything as delicious as that chocolate?" No, I said, nothing.'

Natasha knew little about her maternal family, except that they came from Siberia (her mother died twenty years ago). But a cousin had recently surfaced, from Khabarovsk. She came to Moscow to visit. Natasha's sister travelled to the capital for the reunion from her home outside the city, and they invited the cousin to Natasha's house. The sister went to meet the cousin at the metro.

'How will I recognise her?' she asked in advance.

'You both have mobiles,' said Natasha.

But when the sister looked into the crowd surging like a wave over the top of the metro steps, a face shone out 'like a torch'. She said, 'I saw one of us.'

I met Natasha again in Moscow three or four years after our first encounter. Commenting on recent events in Russia, she cited former prime minister Viktor Chernomyrdin: 'We meant the best, but it turned out as usual.' Natasha noted, as she had before, the immutable aspect of Russian history despite its assorted cataclysms. She dismissed notions of a golden Soviet age when walls were so thick you could not hear your neighbours even if they succumbed to the Russian equivalent of *rar* and hurled themselves into the bath. 'Every fresh initiative produces the same results!' she said cheerfully before she kissed me goodbye at the metro station, waves of faces again surging up the steps.

Many millions of families had lived through trauma like hers, but the indigenous peoples of the far north had been worst off in a ferociously contested field. Thirty-six distinct groups hunt and herd in Russia's Arctic north, from Nenets to Yup'ik, and for them it has always turned out the same, at least since 'Russians' ventured north. In the Soviet period, supply lines to the north were so long and so chaotic that four thousand left-foot gumboots arrived in Yakutia. The right feet never appeared, presumably having defected by limping into Finland. When an indifferent market economy filled the vacuum left by the collapse of communism, it cast Arctic peoples adrift on diminishing floes. Marooned myself once in the Russian Far East, I fell in with a family of marine mammal hunters. They were Chukchi, a people who once hurled stones to down migrating geese and built antler towers as seal-oil lighthouses. They still believe in something, or most of them do, and preserve Chekhov's dignity in their old souls. The natural world, sacred to Chukchi, belongs to a spiritual universe connecting people with the land on which they once depended, and, crucially, with their ancestors. Chukchi don't believe that they own the territory Moscow bureaucrats call *Chukotskiy Avtonomnyy Okrug*. They are stewards. For all of time they have performed ceremonies to thank bowhead whales and walruses for the provision of food, shelter, fuel and sacred objects that embody meaning. In Anadyr I bought a carved bone from an elder. 'These things,' he said, eyes gleaming as he ran his fingers over the bone as a television flickered high in the corner of the café, 'pave our path to the greater world above and below us.' The bone is a walrus baculum, a penis (walruses have the biggest penis bone, relative and absolute, of any mammal). It is forty centimetres long and carved with a totemic human face along with a walrus head and flippers. This object, an emblem of all disenfranchised people, is an ally on my desk and in my head. Marooned again but this time in my own world, one broken in a different way, I think of the dignity encapsulated in this bone.

Uzbeks, disenfranchised in another way but the same one, sold mobile phones outside Moscow's Vykhino metro station. On one rather dismal trip I passed them every day. The women washed their hair in the bathroom of an adjacent McDonald's. I knew that because I went into the McDonald's to use the internet. These Uzbeks too were adrift in a world which didn't cherish them. After my Russian book appeared I developed an interest in the five Central Asian stans that had achieved independence for the first time in their history when the Soviet Union collapsed. The toponyms alone make the heart beat faster: Samarkand, Balkhash, Karagandy. Since the original Great Game, the nineteenth-century tug-of-war between the British and Russian empires, first the US took the baton from Britain, and now China plays a new, Greater Game amid the still-unfolding tragedy of the Taliban's Afghanistan. In 2019 I set out for the first time for Central Asia, heading initially not to political stewpots but the sanctuary of the hammam. In the domed Kunjak in Uzbekistan's holy city of Bukhara I enjoyed a thrashing as women have throughout Transoxiana since the sixteenth century. Later, on the way to Noble Bukhara, the waisted chimneys of a nuclear facility rose from the steppe through the bus window. Then fields of cotton – endless fields. Uzbekistan is the second largest exporter of 'white gold', and Uzbeks have been harvesting it for two millennia. The Soviets ramped up production (90 per cent of the crop went to Russia), diverting so much water to feed thirsty plants that the Aral Sea died, one of the greatest ecological disasters the world has known (so far).

I found the semi-sunken Kunjak among ziggurats of watermelons (pink gold?) in the alleys behind the baked-brick Kalon minaret, a structure built in 1127 also known as the Tower of Death, since various emirs ordered their enemies to be hurled from the top. The masseuses were Uzbek, smiling to reveal rows of Soviet gold crowns as metal buckets clanked on stone. But the owner was a Tajiki-Persian-speaking Tajik, and in the reception area she lounged on a day bed counting out the day's takings in som notes. The economic disparity between Tajik owner and Uzbek workers is characteristic of the double-landlocked country. With 33 million people, Uzbekistan is the most populous of all the former satellites, not just those in Central Asia, and the poorest in per capita GDP. Five changes of alphabet in the twentieth century exacerbated problems of literacy and Covid has further weakened a fragile infrastructure. As for the hammam, etiquette varies, and I once experienced trauma in Tripoli when a female attendant told me off for removing my knickers. I didn't have an interpreter in Bukhara so I acted out the disrobing routine and looked quizzically at Firunza, my masseuse. She replied by acting in turn the business of knicker-removal to indicate that nudity was required (those Libyans are part-timers), then came out with two perfect words in English: 'like baby'. Burqas are illegal in Uzbekistan, and in the changing room women were folding spangled headscarves dyed with pomegranate rind; the younger ones peeled off counterfeit T-shirts advertising 'Channel' and 'Guccy'.

Uzbekistan calls itself a secular Muslim state. Sounds good. But it would be too generous to call the country authoritarian.

The ritual proceeded with a mustard scrub on a stone massage table surely dating from the age of Tamurlane, that Uzbek warrior and national hero who gallops across the five-hundred-som banknote, conquering hand aloft. Apertures in the roof let in light which cast circles on flagstones worn smooth by female feet. Women moved softly in and out filling buckets from low taps. I took the opportunity to talk with Firunza using a mix of her bad-English and my terrible Russian (from 'The Lady with the Little Dog' I knew useful words like inkstand, lorgnette, groyne, grasshopper and of course *beret*) and gestures such as the massage process allows. She had been working at Kunjak for ten years, she was sixty-two, and that day her daughter had come in to help her (she gestured to a cheery figure walloping a client's shoulder blades). Her other daughter worked for a glazier in Moscow, her son in Dubai: remittance money contributes significantly to the economy in all five stans. When I got back to my hotel I mentioned the stone table, now definitely a contemporary of Tamurlane, to my guide. 'I love Timur,' he said, and meant it; he thought of him as a father. It would be difficult to overstate the warrior's role in the national consciousness, even if that role was largely manufactured by the post-independence president: Amir Timur (actually not Uzbek but Turco-Mongol) quickly knocked Marx and Lenin off their perches. He would have approved of the contemporary police state and its wretched human rights record. Tamurlane left five million dead in north India alone. In Uzbekistan, nobody counts the bodies.

It had been a long time since boyfriends had appeared on the travel itinerary. I was not sad. All the men had added something, even if sometimes it was a very small something, and when I did look back, it was with fondness, usually, and sometimes with horror, and quite often with amnesia. Russia suited the unromance of late middle age as there was something graceless about it if human beings appeared in the landscape. But now in Central Asia another form of romance hove into view. A girlfriend, like me in her late fifties, told me that in Borneo young men had approached her in a flirtatious way, and that she noted a culture of economically poor males hooking up with what looked to be money-loaded western women. I had heard about this phenomenon in North Africa, where mature British women on holiday in Tunisia had married and brought husbands back to live in Bolton or Brighton. Sometimes it worked out well for both parties. And why not? But I didn't fancy it. I was too old-fashioned as well as too old, and I couldn't be bothered. So when a handsome brute smiled at me in Bishkek market and offered me a mint tea while taking my wizened hand in his muscular paw – I just couldn't raise the energy.

I was more interested in looking out than in by that stage. I knew myself quite well enough; any still-concealed data could remain concealed. I won't say that Chekhov had succeeded in his aim of teaching me how to live, but 'The Lady with the Little Dog' had taught me how to battle on, and that was enough. And there was a lot to see in the stans. The countries float on the edges of western consciousness as galactic empty spaces – *Kazakhstan*, it is always said, *is the size of western Europe.* Of course they are not empty and their culture could teach our own a good deal. The region is entirely *flat*, the notion goes. But 90 per cent of Kyrgyz territory is mountain – the Tian Shan range slices east to west and you can see its snowy summits beyond the sprawl of the capital, Bishkek. I once crossed the border from Uzbekistan into Kyrgyzstan on foot at the no man's land of the Dostyk post, near the second Kyrgyz city of Osh. Hundreds pressed in both directions, contained behind metal fences. Unlike Uzbekistan, which has oil and gas and a manufacturing industry producing cars and white goods, Kyrgyzstan has no hydrocarbons and no minerals bar a couple of gold mines that spat out the odd nugget. Not long ago a respected economic institute estimated that over 50 per cent of the Kyrgyz economy is black or related to smuggling, the latter often drugs en route from Afghanistan via Tajikistan, a supply-line considerably more efficient than the one involving Soviet gumboots. Across the region our old friends endemic clan-based corruption and a thriving shadow economy queue for attention; drug trafficking is the lifeblood of the Silk Road now.

I fetched up in the Naryn Region about two hundred kilometres from the Chinese border, close to the alpine lake Son Kul, and talked to Baiysh, an elderly Kyrgyz herder. 'My mother,' said Baiysh, 'was a *manaschi*.' She was a teller of tales of Manas, a mythical figure who capered around those parts a thousand years ago, fighting oppressors in order to establish a Kyrgyz homeland among the granite spires. *Manaschis* like Baiysh's mother are professional reciters of the *Manas Dastani*, the epic poem of the hero's nation-building exploits first written down in Persian during the 1790s, though Kyrgyz scholars I met in Bishkek insist it goes back much further. The entire work runs to half a million lines. 'Can you recite a bit?' I asked Baiysh. The levels on my audio recorder leapt into the red as everyone in the yurt shrieked with laughter. 'Only women can do it,' said my interpreter between his own yelps of mirth, 'and you have to go into a trance . . .'

Unlike Uzbekistan, Kyrgyzstan is historically a land of nomads. I was talking to Baiysh in one of his own yurts. He had branched out into adventure tourism twenty years earlier as his country opened up. Beginning with one yurt, he supplemented the income from his flocks of goats and fat-tailed sheep and his herds of horned cattle. Now thirty yurts curled in a line towards the lakeshore. I asked what the most significant business development had been. Through the tunduk, the hole at the canvas apex, I saw kites wheeling. Baiysh thought. 'In 2010,' he said eventually, 'we got seats for the long-drop toilets.' His whole family work at Son Kul during the short June to October season – those, that is, who don't move around with the animals, travelling from distant yurt to distant yurt. Cows grazed the pasture above the treeline, as cattle have for millennia, before independence, before Soviets and even before Manas. At all hours, boys galloped past at full tilt. I had never seen people who have such a seamless relationship with their landscape: human, horse and grassland flowing into one coherent whole. I mentioned this in the yurt. More shrieks of laughter. 'Then you must come to see some *kok boru*, our national sport,' cried Baiysh. So the next day, we went to the horse games.

Two six-a-side teams saddled up on a makeshift pitch. Soon hooves pounded past us few spectators as riders vied to pick up the headless body of a goat, without dismounting and using only their bare hands. The victor was to drop the carcass into the centre of a tyre. That counted as a goal. As the game progressed, the lake changed colour like a paint wheel, shifting from glittering peacock to muted violet. Late that afternoon, Baiysh told me more about farming in Naryn. He seemed weary. 'Some of my neighbours have got small agricultural enterprises off the ground following the hard post-Soviet period,' he said, 'but we herders really struggle to find the costs to vaccinate our animals . . . and there are almost no roads, except the one to China, which Beijing cash recently improved.' China has increasingly involved itself in funding Kyrgyz infrastructure in this strategically crucial region. There is no internet at the farm, but Baiysh said he had it at home, and last year had followed news sites reporting the breakdown of a Chinese-run power plant in Bishkek. 'My brother has six grandchildren living in his flat there,' he said. 'They had no electricity for a week, in sub-zero temperatures.' He was ambivalent about his country's relationship with Beijing, and even more hesitant to give his opinions openly. You can practically see China from Son Yul. 'The most difficult and complicated part of it,' he said, 'is only just beginning.'

A villainous Russia dominated the news agenda while I was writing the last chapters of this book. I often thought of individual Russians I have met over the years: the ones mentioned here, and others who live on in my memories or the notebooks. Sasha and Marina, my homestay landlords in Anadyr in the Russian Far East, whom I sat alongside at their small kitchen table extracting eggs from the roe of an unidentified fish by rubbing beige lumps over the strings of a badminton racquet picked out of landfill. The Uzbek women, now washing their hair somewhere else, as on 8 March 2022 McDonald's closed all 850 of its Russian outlets.[1] My Irkutsk landlady, a woman in her forties with hair dyed the colour of cornflakes, who jumped up when she saw me angling my phone to take a snap. 'I must put on my shoes,' she said, fetching a pair of plastic high-heeled sandals. I said, 'I'm only taking your face.' 'It doesn't matter,' she said. 'I haven't got anyone else to put them on for.' It was not their war. And of course I think of Russians I couldn't know. Some fifty-five thousand children sleep rough in Moscow on any given night. Many have trench foot, a condition not seen since 1915. It was not the children's war either. This week I saw footage of Ukrainian refugees who had escaped into Russian Taganrog, the port on the Sea of Azov where Chekhov was born. 'Where are we supposed to go now?' one woman said to the television journalist, cradling her baby. Chekhov would die of shame for Russia. 'The world,' he wrote, 'perishes not from bandits and fires, but from hatred, hostility, and all these petty squabbles.'

---

1. Later that year some branches reopened under the new name Vkusno & tochka, which translates as the peerless 'Tasty and That's It' – quite the Soviet compromise.

9

'The Bronx? No Thonx'

The fist-shaped borough north of Manhattan is blank on my creased map of New York City, though it represents a place the size of Paris. If the Bronx were a municipality it would be *the seventh largest in the US*. Forty-two square miles and forty-two branches of McDonald's; 1.4 million inhabitants; one bookstore; eighty miles of waterfront; beaches lapped by the western end of Long Island Sound; ghostly Siwanoy pulling herring from the shallow Bronx River with hickory hooks; a vibrant, perpetually refreshed palimpsest of immigrants, from Central Europe, from the Dominican Republic, from everywhere.

I realised, looking back as I have been, that when the sledding gets tough I write off my losses in a chapter. In the Russian book I see it most clearly. It had been a tricky period. Tolstoy said time and age move past you but you feel just the same, and as sixty loomed, I felt, all told, that I did. By 2019 I was ready to write another book, one that looked forward rather than back; a fresh start. At the end of that year, unaware that no fresh starts lay ahead for anyone, I decided to write a book about the Bronx. I was keen to return to the US, and this time planned to centre myself in one manageable-sized place for a year. It would give new perspective to the business of observation. I had read that Bronxites between them speak 160 languages: like a Chekhov story, the borough was a scale model of the planet. I was thrilled with this idea – having been to the world, I would move to the Bronx and let the world come to me. Something of the symmetry of the argument appealed; I see now that I was trying to find a pattern for material which didn't yet exist. I was shortly to enter my seventh decade, Moscow a dot in the rear-view mirror literally and figuratively, though I no longer saw myself in retreat from anything, and the project was to be a kind of reckoning. At the beginning of February 2020 I went on a trial run, planning to stay two months.

I had rented a place in Norwood, a Democrat-voting Latino district at the end of the D-train line. It was an airy one and a half (a double and a single with an unmodernised galley kitchen) in a six-storey 1970s block. In the single bedroom a contraption on plastic legs held a carpeted cylinder on which a cat might sharpen its claws. Where was the cat? At a branch of the New York Public Library just round the corner on 205th I ordered books via the citywide loan system and within a fortnight I had eaten *fahsa* in Little Yemen, *morog polao* in Little Bangladesh, *pho* to go in Little Vietnam and everything in Little Italy, the latter having transitioned into Little Albania-Italy: signs might still be in Italian, but the man stretching the *strascinati* was from Elbasan. Despite the quantity of 'Littles', balkanisation was not in evidence. The Chinatowns were in the less interesting boroughs. I liked that. In fact, I loved the Bronx. Like London, it revealed the enrichment of immigration and gave the lie to the bitterness of 1960s Bristol. I loved coming *home* to my cosy one and a half each night, and considered getting a cat. My travelling life reached maturity on 205th Street.

The geography was not as I had envisaged. A quarter of the borough is green, from the hemlock forest in the Bronx River gorge to the maple groves of Pelham Bay Park where boys perch on fences like birds on a wire. Shaped like a boxer's glove punching low, the Bronx runs at the cuff from Yonkers to New Rochelle, following the Westchester County line. Three ridges rib the west, worn foothills of the Berkshires and Green Mountains. The Hudson flows past the upper west flank and Long Island Sound dissolves off the eastern shore. Peninsulas of salt meadowland with names like Clason's Point and Throggs Neck jut into the East River. I began assembling this information in a red-and-black notebook. The factory had started making them again.

I wandered the streets giving serendipity the upper hand, a mode of travel I increasingly find the only one that makes sense. (It's senseless, until a faint pattern emerges, like a picture developing on celluloid.) I made forays to Riverdale, where JFK and Trotsky lived, the latter, according to his letters, obsessed with the garbage chute; to the beautiful anomaly of City Island; to Edgar Allan Poe's cottage where his twenty-four-year-old wife Virginia died in 1847 in a narrow iron bed and he wrote of 'a despair more dreadful than death'; and to a rally in which Alexandria Ocasio-Cortez, newly elected Democrat Representative for New York's 14th congressional district, addressed her east Bronx constituents. Ocasio-Cortez, whose parents are Puerto Rican, grew up in the Bronx and expressed something of the borough when she spoke of her Latinx heritage. 'To be Puerto Rican,' she said, 'is to be the descendant of . . . African Moors [and] slaves, Taino Indians, Spanish colonisers, Jewish refugees, and likely others. We are all of these things and something else all at once – we are *Boricua*.'

I had, I congratulated myself as I unlocked the door of the Norwood apartment, the cat cylinder watching in silent reproach, discovered one of the most fascinating urban enclaves on the planet. Why, then, did nobody else want to go there? Even friends in Manhattan had never been to the Bronx. 'I look down on it from above,' one said, 'when we drive out of town on the elevated expressway.' Worse, why had the borough attracted a superficially negative image? A clue came in the still-cited phrase, 'The Bronx is burning!'

On the cool night of 12 October 1977, sports commentator Howard Cosell, broadcasting from the bleachers of Yankee Stadium (the Dodgers were whipping the Pinstripes), watched spires of smoke rising from multiple incinerations above the white icing of the cornices. The city had fallen away, and fifty thousand New Yorkers perched in a clamour of dreamlike intensity. The smoke, though, grew darker. Live on ABC, over a crackly mike, Cosell said, 'The Bronx is burning!' Those four words came to define the borough, its past reduced to a phrase which poisoned its present. How had it gone so wrong, this place which had been, in the twenties and thirties, a staging ground for the American Dream? It had reached the tipping point decades before the seventies. Ill-conceived urban renewal projects like Robert Moses'[1] Cross-Bronx Expressway that had destroyed low-population-density areas; lack of landlord investment; drugs; gangs – one commentator wrote of 'an epidemic of abandonment, vandalism and arson', and the South Bronx became 'a national symbol, a disaster area invoked as *the epitome of urban failure*'. The novelist James Baldwin, from Harlem across the bridge, said white flight had a lot to do with it, and he was right about most things. A near-contemporary of Baldwin's, Bronx-raised Don DeLillo, inserted a scene in *Underworld* in which European tourists fly in to stare.[2] The second half of the twentieth century ruined many American cities as the suburbs beckoned, but the Bronx epitomised a nadir of sorts. There the poorest congressional district in the nation had the largest concentration of public housing towers. Every account used the word 'apocalyse'. I began scanning paragraphs to find it before I read them.

---

1. The deadly handprint of public official Robert Moses (1888–1981) is everywhere in the Bronx. 'He thinks he's God,' one critic said. 'But he's only Moses.' People extol Robert A. Caro's multi-volume life of LBJ. Fewer read *The Power Broker: Robert Moses and the Fall of New York* (1974), among the most affecting biographies I have ever read.
2. 'As a little boy,' DeLillo wrote of his childhood, 'I whiled away most of my time pretending to be a baseball announcer on the radio.'

Why does the Bronx remain a catchword for flames literal and metaphorical almost half a century after Cosell spoke his incendiary words? Why did its problems blaze around questions of race long before the murder of George Floyd and BLM and our own ongoing, ruinously late reckoning with the systemic inequality we had fostered? Is the borough a microcosm of that issue in America? In *The Bonfire of the Vanities* Tom Wolfe skewered the Bronx for my generation as irredeemably bad, dark and dangerous, and in *City of Girls*, thirty-two years later, Elizabeth Gilbert does the same for millennials and Generation Z by using showgirl Celia Ray's Bronx background as shorthand for closed-off limitations. Over 40 per cent of Bronxites are African American. The South Bronx is 1 per cent white. Manhattan's Upper East Side, six subway stops away, is 80 per cent white. The Bronx trails last among New York State's sixty-two counties in health indicators, and residents have a life expectancy five years lower than their Manhattan neighbours. The borough hasn't gentrified into a northern Brooklyn or suburbanised into a Queens-joined-to-the-mainland-without-fires. The word 'gentrification' rings through the Bronx story like a Greek chorus but has never evolved into more than just that – a word. Even the borough's name fails to conjure the romance of New York. Eighty-eight years ago Ogden Nash quipped, '*The Bronx? No thonx!*' You couldn't imagine David and Victoria Beckham christening their second son 'Bronx'. Why such an apparently menacing image so close to the sheen of Manhattan? Tom Wolfe's Bruckner Expressway still gleams with iridescent puddles of motor oil as it does in *Bonfire*, but an Oaxacan café on nearby Willis Avenue welcomes undocumented migrants with a green banner saying, CARAVANA MIGRANTE ¡ADELANTE¡. The café was, according to the *New Yorker*, 'a crucible for the resistance in the South Bronx' – resistance to crouching developers with a new Brooklyn in mind who had inevitably christened the district SoBro.

The forays I made extended into the once troubled South Bronx, but I failed to get a purchase on streets which, to a casual observer, seemed permanently unavailable. Life went on behind an invisible veil. Short of ideas, I went online and booked a place on a guided group walk. Required to muster at ten o'clock on a corner of the Grand Concourse, I rode the subway to 145th, inadvertently burrowed into Manhattan's Hamilton Heights and walked east back towards the Bronx, crossing the Harlem on the swing bridge. The traffic was noisy, and the air carried the fruity tang of carbon monoxide. The diminutive Rosa, my guide, was wearing a belted navy raincoat and carrying a clipboard. Bronx-born of Dominican heritage, unusually for a Dominican she had Arctic-blue eyes. She had no other punters that morning. After introductions, we began to walk; it was one of those days when it is summer in the light and winter in the shade, and a circle of clouds momentarily formed a halo round the weak sun. Rosa told me the story of her firm, speaking in a forced loud voice to make herself heard:

I know we have so much to offer in the Bronx – even just what I'm going to show you today. Yet nobody comes here, so nine years ago I launched this company [Bronx Historical Tours], funding it with savings and a part-time job. I had flyers printed and distributed them around Manhattan hotels.

There was hope; a glimpse of a face behind the veil.

For the first scheduled tour I hired a bus and we went round to pick up customers. But nobody came. Nobody. Of course I had to pay for the bus.

We walked up, then down. In 1909 planners modelled the six-lane Grand Concourse on Paris's Champs-Élysées, and the borough has evolved along its five-mile spine. According to one social historian, in the twenties, thirties and forties it was 'a staging ground for the American Dream ... Lauren Bacall, Herman Wouk, Anne Bancroft, Jonas Salk, Armand Hammer, Tony Curtis' ... these were the feet that had trodden Concourse sidewalks. Parts of the thoroughfare stayed grand until the early sixties. 'We used to go up there on Jewish holidays,' a Manhattan acquaintance recalled of the period, 'to celebrate with a school friend and her family. I remember that her apartment, the whole Concourse in fact, seemed terribly sophisticated.' Rosa stopped outside the Concourse Plaza Hotel, once the belle of these Elysian Fields. Builders' netting shrouded it at street level, and the eleven upper floors, an elders' care facility, appeared gloomy and grimy. A drifter lay in front of the main entrance, his head on a horizontal shopping cart. This, then, was the famous landmark which hosted FDR and the Yankees as Tito Puente jammed mambo in the ballroom. What happened? The Jewish community left in the remorseless Bronx progress north, poor management failed to arrest decline, and the Plaza became a welfare hotel. A guest shot the manager. 'I started the firm,' said the undauntable Rosa, hurrying on as if she could redeem the lamentable state of the Plaza by putting distance between herself and it, 'partly because the Bronx has such a terrible image – I wanted to do something about that.' We paused again by the Supreme Court, Moses brandishing the Ten Commandments (the statue is called 'The Majesty of the Law'). 'We used to live down here,' Rosa said, 'in one of these thirties co-ops.' The co-ops still exist, though prison-style gates shield the entrance porches and brick walls shut off the alleys lacing between and behind. Management had stinted on public-access-cleansing, but not on barbed wire. We stood in the shadows. Buttoning the top of her mac, Rosa said,

> There were no gates then. But things changed and we left running out of here, it got so neglected. We moved up to Fordham Manor, then had to run out of there too. Fordham Heights, I tell you, isn't a piece of cake. We kept running, further north, further from the troubles of the South Bronx. Now we're back on the Lower Concourse.

Rosa lived on 167th with her mother and sister. The landlord was selling. The apartment looked down towards Mott Haven, the mottled-brick former ironwork district closest to the perfumed promise of Manhattan, and up to 138th around Third, where developers are erecting plywood walls around semi-razed sites. The word 'Brooklynisation' also rings like a chorus through the Bronx drama. 'We see it creeping from the waterfront,' said Rosa. 'But we don't need upscale residential units.' Activists had sprayed the word 'Colonisers' across a sign for a new Bagel & Barista Café. Posters everywhere advertise legal services to tenants facing eviction. The law was not as majestic at this latitude.

We arrived at the caramel bulk of the Andrew Freedman Home, a building of the little-known monumental-monastic school. Freedman, a businessman and one-time owner of the New York Giants, conceived the project as a 'Home for Poor Millionaires', by which he meant failed entrepreneurs. As a child Rosa used to play in the grounds: 'We dreamed of living here.' She paused, then pointed west, where South Bronx cross-streets dissolved in a fog of exhaust fumes. 'Down there, between Shakespeare and Anderson, people come to see the *Joker* Stairs [where Arthur Fleck walks then dances in the 2019 movie set in a seedy Gotham City]. Activists gather on the roof of the building opposite every day' (she repeated 'every day') 'to throw eggs at tourists getting out of buses to take selfies.' I made a mental note to strike the defenceless Stairs off my itinerary. Tourists apparently imitate Joaquin Phoenix bending back between the railings at the top of the steps. Sometimes they paint clown mouths on with lipstick. 'Protesters,' continued Rosa, 'want the name changed from Joker Stairs – an unofficial name – to Bronx Steps. I certainly won't call it Joker Stairs. I'm not saying gentrification is all bad. It's just that I see it both as a blessing and a curse.'

Not seeming to know what to say next, she asked, 'Know what a Bronx cheer is?' She blew a raspberry to demonstrate, her lips briefly forming a clown mouth. Then she said all her friends had left. Where for? 'Florida. Westchester. New Jersey. The Carolinas. That sucks. Bronx cheers to 'em. We need to be here to fight to rebuild our image.' The tour ended at Yankee Stadium. I had already glimpsed the outfield through a slot in the right-field bleachers from the elevated train approaching 161st Street. Many years ago I went to a game with a prospective boyfriend, a colleague on a magazine (he in New York, I in London). We arrived early, and I remember the white icing of the cornices surrounding the blue-tiered seats, above which, the swain reported, no ball has ever been hit. Whilst 'Yankee' is a global brand and the most famous feature of the Bronx after the troubles, Bronxites resent the Yankee operation. Rosa said,

We can't afford tickets; the average income around 161st is $18,000. Even when I was a child we used to stand outside the old stadium and listen. As for this – [she made a gesture towards the new stadium] community benefits agreements in zoning have a very mixed record of effectiveness.

So even Yankee Stadium reveals the innards of the borough. But the Yankees are part of people's lives. 'I'll never forget the day Thurman Munson died,' Rosa continued. He was a catcher from Ohio. 'It was in a plane crash. All the kids were crying.' The new fifty-two-thousand-seater stadium, raised on heavily used parkland with dizzying public subsidies, opened in 2009, but resentment seeded the day the decision to build was announced. 'Complaining does us no good,' Rosa said. Resentment was old. Forty years before the new Yankee stadium unlocked its turnstiles, Bronxites complained about the portrayal of the borough in *Fort Apache, The Bronx*, in which Paul Newman stars as a patrolman serving a drug-crazed patch, pausing to deliver a baby. The protestors did wrench concessions out of Time-Life Films, but they didn't add up to much.

The tour ended, and Rosa grew smaller as she walked down the steps of the 161st Street subway. I knew that her resonant voice would speak again on the page. I had a purchase on the material. That was enough. More than enough: a glimpse behind the veil, a brief lift of the curtain, a fleeting view through a keyhole in a fortress wall – these are the moments that make it worthwhile.

Over the next weeks I haunted La Morada, the Oaxacan restaurant I mentioned next to a funeral home on Willis Avenue. It gave a face to a more positive version of the 'activism' to which Rosa referred. A sign on the front door promised, REFUGEES WELCOME. Inside, Natalia, the patronne, wearing a pinny embroidered with sunflowers, ferried domes of *mole poblano* from the open-plan kitchen. She and her husband, Antonio, were farmers in northwest Oaxaca once; they had been running La Morada for ten years. Besides providing cheap food, they organised legal rights seminars for immigrants, and *indocumentados* gathered round the wooden tables at weekly information sessions. The *mole* had a bosky undertow of dark chocolate, and chicken *flautas* languished in a luscious pool of avocado, tomato and chilli. Natalia came to collect my plate. A man was whistling in the kitchen.

'*Muy rico*,' I said of the *mole poblano*. 'What's your secret?'

She laughed. 'Three types of chilli: ancho, guajillo and pasilla.'

Natalia took out a handkerchief and blew her nose. Grateful to take the weight off her feet, she sat alongside me to rattle off the ingredients of *mole poblano*, a list so interminable that I made a note, adding to the crossed-off Joker Stairs, never to attempt this delicious dish. Antonio arrived with a glass jug clinking with ice to refill my *agua fresca*. The green migrants banner I had observed hung behind me. The other side, of course, was watching. Last year, undercover police arrested one of Natalia and Antonio's daughters on unspecified charges as she waited table, later releasing her without indictment. But there was spirit on Willis Avenue. When I left one night, Natalia and a helper sat facing one another, de-cloving garlic bulbs from a pink washing-up bowl set on the table between them. Natalia had her back to me, but the helper waved a plastic-gloved hand through the smeary plate glass window as I picked my way among the yellow-lit Willis puddles.

Immigrants to the Bronx in the first half of the twentieth century were almost always in second flight: the first from Europe, the second from the garment factories of the Lower East Side. The novelist E. L. Doctorow's grandparents, fleeing pogroms in Russia, made the pilgrimage from Ellis Island to the Bronx via Manhattan. In the 1985 novel *World's Fair*, which hews closely to an autobiography, the narrator Edgar (born a few months before the real Edgar) remembers bearded men from Manhattan yeshivas knocking on the door 'with coin boxes and letters of credentials'. Rose, the narrator's mother, says, 'When my father brought us to the Bronx when I was a little girl, he didn't know

the whole Lower East Side would follow.' I followed myself, looking for the house where Doctorow (1931–2015) had grown up
during the Depression. From the Concourse I walked up East
173rd and turned into the northern half of Eastburn Avenue. The
East Tremont area was sucking in refugees from New York slums
at the time these houses rose from the coastal basalt. Green and
red buds were already plump on the branches of Norway maples
when I photographed number 1658 and the cream curtains in its
casements. A man was washing his car. The air smelled of solvent
and detergent. Wiping a soapy gauntlet on his sleeve, the man
asked, 'What's up?' I told him that Doctorow, a writer, grew up in
that house. 'Is he dead?' I said he was. 'He and his family,' I said,
struggling to think of anything to say about Doctorow, whom I
consider a major minor writer but didn't want to get into it, 'had
to move from the first floor to the second in the 1930s, and he
recorded that from the window in his new room he could see a
crescent of grass in Claremont Park.' There was a pause. 'I can
still see that,' the man said, returning to his fender.

In the first talking picture, *The Jazz Singer*, released in 1927, Al
Jolson promises his mother that when he makes good they will
move from the Lower East Side to the Bronx. North of 132nd:
sunlit uplands. The Bronx offered a glimpse of a better world.
Doctorow describes wet sawdust on the floor of Irving's fish
store, a place which 'had a kind of swimming-pool atmosphere',
the travelling farm exhibit in Claremont Park, and his school on
173rd, a watercolour of FDR hanging above the blackboard. And
of course he describes the stoops, the front steps that feature in
every account, from his own *Billy Bathgate* to Robert de Niro's *A
Bronx Tale*. It was where you sat to contemplate your Bronx world.
Doctorow was named after Edgar Allan Poe, whose cottage in
what is now Fordham was not far from Eastburn Avenue. 'Did
you and Dad realise,' the adult Doctorow once asked his mother,
'you named me after an alcoholic, drug-addicted, delusional
paranoiac with strong necrophiliac tendencies?' Poe and his

household were so poor they ate turnips planted for cattle. To maintain spirits, they kept a tortoiseshell cat called Catterina and birds that hung in cages from fruit trees in the garden. It worked: Poe's tenancy in the Bronx was productive, yielding, among other poems, 'Annabel Lee'. *It was many and many a year ago,/In a kingdom by the sea,/That a maiden there lived whom you may know/By the name of Annabel Lee.* When the maiden dies, the poet lies on her grave, as Poe did on his wife's grave, hence Doctorow's 'necrophiliac' snipe. Poe had already mastered the short story, half a century before Chekhov. Visiting his cottage today, marooned on a kind of traffic island off the Concourse, I imagined not Annabel but Maria Sergeyevna doppelgangers in berets leading little white Pomeranians along the streets around Jerome.

My own street, the long northwest Decatur, decanted at the northern end onto a dicey stretch of East Gun Hill Road before vanishing into Woodlawn Cemetery, and, going the other way, ran southwest for fourteen blocks to hit the East Fordham Road. In the first decade of the twentieth century, when movies first flickered in New York City, Edison Studios built a concrete, glass and steel facility on the corner of Decatur and Oliver Place. Half a century later, in a single year police made four hundred drugs arrests *at one Decatur address*. On the main shopping drag in Norwood, along Bainbridge and 204th, bodegas called themselves delis and other small stores offered services for sending money to foreign lands or certifying divorce papers. Close to Curry & Kebab, already a hangout of mine, launderettes clunked through their cycles day and night and workers trundled past my building pushing square metal trolleys freighted with monstrous sacks of laundry. An improbable number of barbers, all permanently busy, played hip hop good and loud. Customers still waited at eight in the evening.

'It's not feasible,' I said to Warren, the amiable Super in my block, referring to the high barber frequency. He was cleaning the lobby windows.

'It's a Latino thing,' he said. A spry African American in his seventies, Warren was six foot three, and had been on duty at Decatur for twelve years. My landlord had advised me that he was unfailingly helpful, and he was. Warren lived with his wife, two daughters and two young granddaughters in a small apartment on the ground floor, opposite the laundry and boiler rooms. Once I saw him dragging a case of old-fashioned computer paper with perforated edges through his front door. 'Picked out your coffee spot?' he always asked cheerily in the morning. One day, after he had got used to me, I sat on one of the benches in the lobby while he cleaned the floor using a mop and a red tin bucket. I realised I was on an indoor stoop.

'I was a loyal government servant for ten years,' Warren said when I enquired what he had done pre-Decatur. 'In uniform.'

'Where did you serve?' I asked.

'Vietnam,' he said. It seemed he wasn't going to fill the pause that quivered like cumulonimbus over the wet floor. Eventually Warren swivelled to a different part of the lobby and talked about his twenty years delivering fruit and veg in the hood; he had enjoyed it until his eyesight deteriorated – it had been damaged in the military – and he could no longer drive. We enjoyed many conversations during my tenure, and I was always pleased to see Warren. He never asked me anything about myself. I got the idea that something had shut down, perhaps in a good way, and that he lived in the moment.

Morning and afternoon, neatly dressed Puerto Rican children shuttled to and from school holding parental hands. Norwood had not always been Latino and Latina. A Northern Irish tribe pitched up from the Depression onwards and dominated the neighbourhood, especially in the sixties and seventies. There was even a Little Belfast. They had mostly moved on – I never saw an Irishman at the bar of The Little Tinker on my part of Webster – but Italians had stayed. Bakeries like Delillo on East 187th founded by subway-builders from Puglia and what was then Abruzzi e Molise continued a lively trade in *cannoli* and a man in an indoor market on Arthur Avenue rolled cigars as they used to in the backstreets of Naples and the slums of the Lower East Side. A butcher displayed rolls of tripe like carpets in the refrigerated unit in front of his shop, and at the Calabria Pork Store products swung from the ceiling, covering the polystyrene tiles in what the proprietor advertised as a 'salami chandelier'. Yet they spoke Spanish in Little Italy-Albania. At an Ash Wednesday mass at Our Lady of Mount Carmel, a Romanesque Revival church raised by Milanese labourers and decorated with martyrs shedding lozenges of blood, two front rows of black-swaddled devotees chanted in Italian before an African American priest said the liturgy. But the service was in Spanish. *A scale model of the world.*

Nobody lived on the sidewalk in Norwood like they did on Jerome in nearby Kingsbridge. (The avenue was named after Lady Randolph Churchill's family – she was born Jennie Jerome.) There people asked me for a dollar, especially on Sundays, when sunlight fell through the stained-glass Tiffany windows of St James's Church from both sides at once. My landlord, a cat-owning, Connecticut-based playwright, said on the phone that he had picked the neighbourhood because it was safe. But it falls within New York City's 52nd precinct, of which *The New York Times* reported, 'Few here are foolish enough to sleep with their bed by a window.' Fortunately I only read that after I got home. I could see Jeromes and Churchills had moved on. Investigations revealed that the worst incidents of crime, most of them gang-related, take place in the Fordham neighbourhood near Tiebout Avenue, between Webster and the Concourse at the 182nd level. I often walked there. So had the teenage Stanley Kubrick, who wandered round in the forties with a Graflex in a paper bag, the lens sticking out of a hole at the front. Kubrick left the top of the bag open and looked down into the viewfinder. His father was a doctor, and the family lived in their own house on Clinton

Avenue, not far from the Edison Studio I mentioned. (Before the studio burned down it produced the first film adaptation of *Frankenstein*, foreshadowing Kubrick's Hollywood debut by just forty-six years. You can see a few strobe-like frames on YouTube. Forty-six years seemed a short time for an industry to progress from the black-and-white and silent *Frankenstein* to the noir and noisy *The Killing*.) The young Stanley sold his first photograph to *Look* magazine in 1945. It showed a Bronx news vendor framed by posters announcing the death of FDR. I said in Chapter 5 that for more than two centuries Cajuns have walked a tightrope between assimilation and independence. The opposite applied for those who streamed up from the Lower East Side, like Kubrick's forebears. They wanted in.

Even in February warm breezes announced the arrival of spring, but outdoor swimming pools stood empty. The indoor YMCA in Castle Hill took more than an hour to reach on two buses. Scribbled observations fill the pages of the notebook I pressed to my knee during those journeys. Along Starling Avenue wind pawed laundry strung between casements, and in an upper floor a man sat in his undershirt, his face lit by guttering fluorescent images. At street level a woman picked oranges from a crate balanced on other crates outside Ittadi Bazaar, her left hand gripping the bar of a stroller. The Olympic-sized bath at the Y overlooked the East River, Whitestone Bridge and Queens (Queenie – that would go nicely with Brooklyn for the Beckhams). Waiting for the bus home in an almost deserted Zerega Avenue, planes flew low towards La Guardia and JFK, cats flickered in the shadows, and the Bronx petered out.

I met blue-eyed Rosa again in March to walk in the Botanical Gardens near my apartment. She was a paid-up Friend of the Gardens. I had told her about my book project. 'I've never heard,' she said, 'of anyone from over there being interested in us.' The weather was changing like a wheel at that time. Resentment grew in the gardens along with the orchids in the conservatory. City

Hall had renamed it the *New York* Botanical Garden, as opposed to the Bronx Botanical Garden, Rosa said, this time wearing a baggy nylon sports jacket, 'to get rid of the word "Bronx". They thought "Bronx" made it seem less important.' A sense of inferiority was embedded in the borough's DNA. When Doctorow's Billy Bathgate goes to work for his criminal masters in a midtown nightclub he says, 'Manhattan was only what the Bronx wanted to be.' The Bronx Zoo though, on land contiguous with the garden, had gone the other way, nomenclature-wise – it was originally called the New York Zoological Park. Before the fires, zoo news dominated coverage of the borough in the British press, whether noting the birth of a chimpanzee or the arrival of a polar bear Admiral Peary's skipper had conveyed from the Arctic, or the time in 1934 when two boys stole a nine-foot Bahamian boa and six other snakes and took them to school for show-and-tell. By 1998 London newspapers had lost interest in the zoo, focusing instead on UK-related events in the Bronx. These were few. But that year a Bronx-born teacher who had moved to the UK to work in a school in the troubled Paulsgrove area of Portsmouth took his pupils on a school trip to the Bronx 'to show them the effects of serious deprivation'. The borough, announced a *Times* reporter, was 'a global symbol of crime, decay, violence and social meltdown' (this was twenty-one years after the Bronx burned). Jim Robertson, thirty-eight, the teacher, told the paper that his quality of life had improved since he moved from the Bronx to Hampshire, presenting as evidence the fact that he was 'the first person in our family to own a lawnmower'.

Another snake catapulted the zoo into the British press in 2010 when a two-foot killer Egyptian cobra escaped. 'Snakes are the great Houdinis of the reptile world,' said its sheepish keeper. In 1906, his predecessors had displayed Mbuti bushman Ota Benga in the primate cages. Slavers in the Congo Free State had sold Benga for a pound of salt and a bolt of cloth to an American keen to acquire exhibits to amuse the public. The slave trade had ended on both coasts, but the white man found a way. The African was free to walk around zoo grounds when not on display. Leopold II's militia had killed Benga's wife and children. He wanted to go home to the Kasai River. When he was about twenty-seven a well-wisher paid for him to leave the monkey cage and settle in Lynchburg, Virginia, where he briefly worked in a tobacco factory and had his sharpened teeth capped. In 1916 Benga borrowed a pistol, built a pyre, hammered the caps off his teeth and shot himself.

In one of the streets of cartwheeling plastic bags in the South Bronx I met Noëlle Santos. Operating with the ambitious slogan, 'Let's Bring a Goddam Bookstore to the Bronx', Santos had done just that. Barnes & Noble had closed the doors on its sole Bronx bookstore in 2016; three years later, Bronxite Santos raised the finance to open the Lit. Bar in Alexander Avenue. BOOKSTORE & CHILL, the metal sign over the door said. Santos, then thirty-three, is the daughter of a Puerto Rican mother and an African American father whom she didn't know. On my first visit to the Lit. Bar, Santos, raised in Soundview (a neighbourhood described with mixed affection by Supreme Court judge Justice Sonia Sotomayor in her autobiography), was stationed at the till wearing a fitted black top and gold hoop earrings. Up till her early twenties, she said, 'All I wanted to do was get away from the Bronx.' She thinks differently now, her mind changed by the potential she knows is there. Eight years previously, working for an IT company in Manhattan, she took a course in running a bookstore. Then she won a statewide competition for entrepreneurs and crowdfunded $200,000 on Indiegogo. A volume entitled *Stop Telling Women to Smile* stood to attention on the front table. In the bar at the back, next to two sofas, I asked the friendly bartender for a coffee. As yet, she explained, the bookstore stocked only wine: would I care for a glass?

*

A thousand years ago the Siwanoy people, a branch of the Algonquin Lenape, settled the tidal inlets and open shorelines of Eastchester and Pelham Bay. Pyramids were rising at that time on the banks of the Mississippi. Siwanoy paddled elm-bark canoes and tulip-tree dugouts that carried twenty. We know this from excavated middens, and from them we also know about the deer and mushrooms and sweet mulberries they ate, and about the pottery they crafted. Archaeologists even found ash pits under East Tremont where Doctorow had excavated his own small slice of the past. In the 1620s, as descendants of the first Siwanoy continued to paddle and carve, the Netherlands claimed the east coast land that includes the Bronx, placing it under the control of the Dutch West India Company. Soon Dutch colonists handled much of the Virginia tobacco crop, shipping it from New to Old Amsterdam. Meanwhile 190 miles up the Atlantic coast in Massachusetts Bay, Puritan colonists renamed their Trimountaine settlement 'Boston'. Word of prosperity in New Amsterdam spread along the wharves of the old country. More settlers arrived in the touted land of plenty after a three-month voyage, four if the winds failed. Jonas Bronck was among them. He came from Amsterdam in 1639 (most sources identify him as a Lutheran Swede, but recent, unpublished research links him with the Faroes), bringing a wife, a posse of indentured farmhands and a small library. In a state of high excitement, Bronck was soon writing home to a friend that . . .

The invisible hand of the Almighty Father surely guided me to this beautiful country, a land covered with virgin forest and unlimited opportunities. It is a veritable paradise and needs but the industrious hand of man to make it the finest and most beautiful region in the world.

Bronck sowed a five-hundred-acre grain and tobacco plantation between pin oak and rocky knuckles. The farm, soon known to other colonists as Bronck's land, spread southwest over today's Mott Haven. History is written by the conquerors, and seventeenth-century century Dutch tracts have told their version of the New York Lenape story during Bronck's tenure. There were massacres on both sides. The Company ordered retrenchment to the tip of Manhattan, where a fort offered protection. On the whole, however, settlers squabbled with each other at least as much as they did with the indigenous people. Meanwhile Britain had never recognised Dutch claims and in March 1664 the restored Charles II granted New Netherland and New Amsterdam to his brother the Duke of York, who took the colony without a shot being fired. In the ensuing negotiations, in which New Amsterdam became New York, Britain offered the settlement back to the Dutch in exchange for their sugar factories in Suriname. The Dutch declined, history followed another fork in the road, and I wonder, reader, if you can find Suriname on a map.

Today, greasy sleeping bags and dented cardboard boxes litter the underpass of Pelham Bay Park subway station, terminus of the 6 train. I crossed three main roads to reach the park where Siwanoy once made pots. The colonisers took all the land for themselves (of course they did) even as America was fighting its costly Indian Wars. If one feature has dominated my more or less random travels over four decades and seven continents, it is minority groups under pressure or gone. The hickory and maple forest of the Bronx, precious and sacred to Siwanoy whose names we will never know, attracted merchants and industrialists who built mansions close to Long Island Sound, a kind of nineteenth-century Hamptons. The screw of history turned again though. In the 1930s, officers at City Hall ordered the demolition of the fine homes in the park to promote 'development'. The past was the enemy all over the US then, so Americans wished to tear it down. Just one house remained in Pelham Bay Park – the Bartow-Pell Mansion, a cut-stone Greek Revival seat where Robert Bartow and family celebrated their first Pelham Bay Thanksgiving in December 1843. (At that time, individual states decided when, and if, to mark Thanksgiving.) Dusk was falling when I left the mansion to walk back to the subway. It was the time, were we in India, that would be called 'cow dust'. But we were in the Bronx, and people were not bringing cattle home: they were travelling on buses from poorly paid jobs, unable to get a seat even on a double Bx141. Weary despair hung like a pall over the huddles swinging from looped plastic straps stapled to the buses' cross-hatched ceilings. Union appeals to raise the minimum wage had recently failed. Progress? In 1950 forty-seven-year-old Marvel Cooke, an African American labour activist from the Bronx, wrote a set of investigative reports for the New York-based radical newspaper *The Daily Compass*. She was describing her experiences in the thirties, the time when City Hall planners were smashing down mansions in Pelham Bay Park.

> I was part of the 'paper bag brigade', waiting patiently in front of Woolworth's on 170th St, between Jerome and Walton Aves, for someone to 'buy' me for an hour or two, or, if I were lucky, for a day. That is the Bronx Slave Market, where Negro women wait, in rain or shine, in bitter cold or under broiling sun, to be hired by local housewives looking for bargains in human labor. Today, [1950] Slave Markets are starting up again in far-flung sections of the city. As yet, they are pallid replicas of the Depression mode: but as unemployment increases, as more and more Negro women are thrown out of work and there is less and less money earmarked for full-time household workers, the markets threaten to spread as they did in the middle 30s, when it was estimated there were 20 to 30 in The Bronx alone.

I got one of the last flights out, in the third week of March, 2020. Newsfeeds ran continuous footage of overflowing airport concourses as frightened people battled to get home or battled to get tested before they were allowed home, or both. Covid lent the Bronx project dire substance and vital resonance. Conversations with friends continued over those awful months via WhatsApp. Of the five boroughs, the Bronx had the highest rate of infection, hospitalisation, and death. The day after I landed at Heathrow, *The New York Times* ran a story by a veteran correspondent, a man who had been at Darfur. He had access to the 'hot' [infectious] wards at Weiler and Montefiore hospitals, the latter minutes from the Norwood flat, and filmed a segment there. Medics wore welding masks and ski goggles. Nurses intubated through holes in a plastic box placed over a patient's head. A technician tied down the hands of a twenty-three year old to prevent her ripping out her ventilating pipe in hallucinatory panic. An assistant physician performed CPR amid a symphony of bleeping bedside alarms, each chest compression unleashing a brew of virus. A red telephone rang, announcing the arrival of another desperately ill patient.

It was not a hospital. It was hell.

Outside the window, amid triage tents in the streets I had walked, ambulances backed up. 'Death here has no dignity,' concluded the reporter. 'The sick can't even see their nurse's eyes.'

Co-op City dominated the view from a bus I had taken regularly. It was the largest cooperative housing project in America, perhaps in the world, a milky-coffee, galactic cluster built on swampland chilled by a wind coming off the Hutchinson River. Now sixty thousand people were self-isolating above its breezy walkways. More than 60 per cent were African American, and 27 per cent Latinx. By May the media was proclaiming the borough 'New York's virus hot spot'. Norwood actually featured in one report, accompanied by a description of Covid marching 'from apartment block to apartment block'. Every day I thought of its arrival in Decatur, for it surely had arrived there, and of the amiable Warren, and his apartment crowded with grandchildren.

The Bronx burned again.

Early in Covid's first summer many stories focused on the forty-four-storey River Park Towers in Morris Heights. You can see it from the bluff of Bronx Community College. I had had my eye on the outdoor swimming pool in nearby Roberto Clemente State Park, planning a plunge as soon as it reopened for the season. The seventies-built towers, known as River Park Residences, consist of two L-shaped red-brick constructions housing 1,654 families, most of whom receive rent subsidies. Green benches mark out the few hundred yards between the two towers. The whole complex appears neither as shoddy nor as litter-strewn as high-density estates in Europe. Conditions inside the blocks, however, were parlous even before coronavirus. A Bronx friend whose cousin lived on the twenty-fourth floor reported that she could wait up to an hour for a lift, which often broke down. At rush hour, residents jostled in hallways like commuters on a Japanese subway platform. On 27 May 2020 *The New York Times* reported that as many as a hundred River residents had coronavirus. 'But no one knows for sure,' the report went on, 'since the leader of the tenants' association died from Covid-19 in April.' The cousin said, 'There always seems to be an ambulance outfront.' Then she died. The Ebola scare in the 2010s had amplified cultural and racial prejudices in the Bronx, drawing on what one professor of medicine and the humanities called 'an "othering" of disease that recalled the HIV panic'. The situation in River Park, according to one journalist inured to tragedy, 'reflects a legacy of institutionalised racism, poverty, cramped housing and chronic health problems'. So disease really is a metaphor in the carriages of the D-train.

A Bronx economic development official warned that up to half the borough's restaurants may never reopen. I recalled the *tilapia bhuna* served by the owner's daughter at Curry & Kebab. Its website indicated it was not even offering a reduced takeout service. 'We always pay the highest price,' a friend lamented in a text. The crisis had exacerbated disparities in existence since Doctorow's Billy Bathgate reeled at the affluence of a Manhattan nightclub. I watched television pictures of municipal workers raising the unknown dead on the Bronx's Hart Island, hitherto entombed in the largest mass grave in the US, in order to make room for the newly expired. Rosa, whose landlord had paused the sale of her building due to Covid, told me on the phone that Woodlawn Cemetery was the only place where you could not hear the klaxon of doom. When I watched, via Zoom, a read-through of my Norwood landlord's latest play, it seemed to circle around themes of exclusion and loss.

Most often, I thought of a day I had spent on City Island at the western end of Long Island Sound. A causeway there joins the finger-shaped island to the rest of the Bronx. A touch of New England clings to the pastel clapboard houses that face the water off both sides of the single island artery; Sidney Lumet picked one to shoot the 1962 film *Long Day's Journey into Night* – actually set in coastal Connecticut – starring Katharine Hepburn as a morphine addict (in the opening scene, the disintegrating family sits on the porch talking about the foghorn). On that day I ate shrimp cocktail from a plastic tray at Johnny's Reef on Belden Point. It was crowded with laughing customers and the windows looked onto sparkling teal water where British gunships had anchored during the Revolutionary War. Where were the customers now?

But during the 'rona the spirit of the Bronx didn't retreat. As the world complained or worried or died, Samelys López had been cruising South Bronx streets delivering food boxes and a message: *vote for me.* Aged forty, hair scraped into a bun, López grew up in the Bronx's homeless shelter system. She was running for Congress in NY-15, a Democratic stronghold which had returned Hillary Clinton with 94 per cent of the vote. It is the most left-leaning seat in the country, and the poorest congressional district. López's living room doubled as a campaign office. It was López, though, whom AOC backed, and so did Bernie

Sanders. López's 2020 platform centred on housing as a human right, universal healthcare and a Green New Deal. She lost. But she got traction. Meanwhile at the Lit. Bar, Santos had collated a booklist she called Dear White People, and when she put it online it went viral. They still believed in Jonas Bronck's veritable paradise and so did I.[1]

'Only connect.' I clung to Forster's advice more than ever, now every advancing stranger was a biohazard. Other writers said the same: it was the only thing we could do. It turned out though that the connection business was subject to tighter controls than it was when Forster brought Aziz and Fielding together in Chandrapore. In the last month of 2020 a new team at my New York publisher decided I wasn't the right type to write about the Bronx. (They would have said 'No thonx' if they knew anything about Ogden Nash.) Everyone involved, including my professional advisors, capitulated not overnight but that same afternoon. A policeman had murdered George Floyd in May; change was long overdue. The small irony that I had sought to let disenfranchised voices speak to facilitate understanding had become collateral damage. Rosa, Warren, Noëlle, the undocumented migrants of Willis Avenue; Marvel Cooke waiting at the slave market in the thirties; Ota Benga in a cage at the zoo in the 1910s; the Siwanoy fashioning potsherds under the firehouse in East Tremont – their voices were not to be heard. Of all places, the Bronx was multicultural. I thought it had something to say. I was happy to leave the saying up to a person of colour, but it was hard, according to them, to get anyone to listen. I did not fancy myself as a white saviour and was not trying to speak for them or co-opt their voices – that, as I was acutely aware Ursula Le Guin had said of her own portrayal of characters of colour, 'would be an act of extreme arrogance'. I remembered Rosa's comment: 'I've never heard of anyone from

---

1. Even Nash eventually recanted. In 1964 he declared in print: *I wrote those lines, 'The Bronx? No thonx';/I shudder to confess them./Now I'm an older, wiser man/I cry, 'The Bronx? God bless them!'*

over there [Europe] being interested in us.' And now we weren't allowed to be interested. Rosa had stayed when her friends fled, she said, because she wanted 'to rebuild the Bronx image'. Couldn't I have helped, in my modest way? I had taken a cue, or tried to, from Chimamanda Adichie. 'If you don't understand,' she wrote in *Americanah*, 'ask questions. If you're uncomfortable about asking questions, say you are uncomfortable about asking questions and then ask anyway. It's easy to tell when a question is coming from a good place. Then listen some more. Sometimes people just want to feel heard.' I did think writers of colour had the authenticity I lacked in recording these Bronx lives. But they had other books to write and other livings to earn. So once again, voices fall silent. It seemed contrary to the spirit of the borough.

Practically, it didn't make much difference to me. But I was sad. Was I also bitter about the money I had spent on the Bronx idea, moving continents and so on? No, I don't think a writer can be, if she is able to pay the bills. Everything that goes in comes out somewhere, even if you can't recognise it by the time it reappears. But something about the way the Bronx and me ended sounded off. I said at the beginning of the chapter that I foresaw the project as a kind of reckoning. And it was, though not in ways I had imagined.

10

The Saddest Pleasure

Zanzibar

Maybe it is sad to travel and learn that life's a bitch
in Nairobi and Manaus and Tokyo and Sydney.

MORITZ THOMSEN, *The Saddest Pleasure*

'This,' he said, gesturing at a low stone bench integrated into the mansion wall, 'is a *baraza* – most important word in Kiswahili. For serious talking.' Opposite, two women of about my age bent together on their own *baraza*, heads almost touching. Hot afternoon light gilled down the alley. 'Where we sort out the world!' said my friend, shaking a liquorice leaf from the hem of his kanzu. He was describing a stoop, Zanzibar-style.

Twenty-one months had elapsed since I'd left the friendly Bronx stoops in a hurry. Like everyone else, I stayed home during what turned out to be a Biblical plague. When I could finally get out, I picked Zanzibar because it wasn't on a red list; because I had never been; and because its own past was layered and nuanced and so, I had understood through writing this book, was my own. (Also, Tanzania, to which Zanzibar belongs, was sixty years old, like me.) Geological strata make a rock, just as waves of history make a place and a storied past makes a person. As I have said, on the downward catwalk to old age the past had become more important – Martin Amis's whole new empire. As I looked at episodes from my life on the road, the interlapping and overlayering of topographical history drew me in.

As the day of departure approached, I began to wake early, already alert with anxiety. The arrival of Omicron had done nothing to allay worry. But still, I felt a victim of Stockholm syndrome, identifying with my viral captor and fearful of venturing beyond the familiar. I had absorbed the *campanilismo* I attributed earlier to my grandmother and her generation – the inclination never to venture beyond the range of the village bells. Only twenty-one months grounded, and I was uncertain if I still had the ability to travel alone. On the day before my flights to Zanzibar I found myself standing in my office with a hairbrush in my hand, about to place it in my backpack. A hairbrush? I had never taken one abroad in my life. Had I reached a new point of degeneration? Would it be a travel hairdryer next? I knew such a thing existed as I had seen one in the Argos catalogue.[1] On the day I left, the sky was a tender blue through the window of the Thameslink train to Gatwick, the dome of St Paul's peeping playfully between buildings old and new, and shadows falling long across the platform at Upper Norwood.

Jet lag was harder than it used to be, but one wonders what isn't. All through the first day, having landed at Abeid Amani Karume just after dawn, I observed the world from behind a veil. But on the second morning I took my place at Livingstone's, a beachfront café on the very spot – I'm going to believe it – where the fervent anti-East African slave trade campaigner prepared for his last expedition, the one after which his heart, literally, stayed

---

1. What a tragedy when Argos discontinued the printed catalogue in 2020. I greatly enjoyed leafing through a thousand pages of items I didn't need. Scrolling doesn't do the job. Another sign of age, I suspect.

in Africa, buried under a mpundu tree by his loyal retainers. From that vantage point I watched girls splashing in the water, hijab swimming hoods framing smiling faces. A man patrolled the sand selling individual eggs from a stack of green plastic trays, and hump-backed Brahman cows lowed in tune with the elephant grief of the Dar ferry. Zanzibar was reborn on the beach every day. The girls retrieved their abayas from a coral shelf in an embrasure on the beach at a place carpeted with plastic bottles and coconut husks, while boys and young men filtered onto the sand, kicking a football, counting one another's press-ups, skipping with a rope. On a propped-up blackboard, Livingstone's displayed an impressive range of coffees from ristretto to cappuccino, all spelled correctly and made from a tin bowl of jammy Nescafé powder. The coffee tasted good. I was far from home and hairbrushes, where I like to be, and a blessing of a sort came down from the missionary Livingstone. I tried not to imagine whether I would strain to recapture the moment in another twenty-one months, if the walls closed in again. Given the baleful pandemic suffering, impossible to ignore in the Bronx and every other place, I realised in Zanzibar that from now on the still sad music of humanity would play everywhere. *For I have learned*, Wordsworth wrote in the matchless 'Lines Written a Few Miles above Tintern Abbey', *To look on nature not as in the hour/Of thoughtless youth, but hearing oftentimes/The still, sad music of humanity.*

I had taken an Airbnb in Mji Mkongwe, the crooked-laned part of the capital, known as Stone Town in English, and that night spotted a red carpet outside the fort, hand sanitisers on posts guarding the mottled wall like sentries. The air was steaming, and a firewoman fanned herself with her helmet as she leaned against a precautionary engine. I followed several thousand unmasked Zanzibaris over the egalitarian carpet and into Ngome Kongwe, the early-eighteenth-century Old Fort, for the premier of the first film made in Kiswahili and shot in Zanzibar. In the amphitheatre,

curved stone benches climbed behind two front rows of cloth-
swathed VIP seats, and in the gloaming, under a titanic screen,
the lush strings and percussive drums of a six-man, keffiyeh-
hat-wearing taarab ensemble accompanied a female singer in a
long, striped robe. Taarab is an Arabic word meaning 'have joy
with music', and that is what those players did. Before the film
began, though, the speeches. Silence settled over the otherwise
restive crowd as the second vice president, wearing a kanzu and
western suit jacket, spoke for forty-five minutes. The sun finally
set, presumably out of boredom. All speeches are long if you don't
understand Kiswahili, though a friend later told me that this had
been a short one. The crowd remained silent, a testament to their
respect, and perhaps to the Tanzanian government's approach
to free speech. The 2021 film, *Vuta N'Kuvute*, or *Tug of War*, was
the country's inaugural period drama and its first feature film to
screen at the Toronto International Film Festival (TIFF). The premier was an offshoot of the Zanzibar International Film Festival
(ZIFF), an annual celebration touted as the largest cultural
event in East Africa. *Vuta N'Kuvute* had already won international awards and secured continental distribution with Africa's
Multichoice. The author of the novel on which it was based, Shafi
Adam Shafi, sat behind towers of shrink-wrapped paperbacks at
the rear of the tiered seating, smiling as he signed. Set in 1954
in the dying days of British colonial rule, book and film tell the
story of a love affair between Yasmin, a young Indian Zanzibari
fleeing an abusive arranged marriage (her husband beats her for
dancing), and Denge, a communist revolutionary. Period detail
included Bakelite telephones, policemen in red fedoras and a
lamplighter who wobbles through Stone Town lanes using a pole
to flick the switches of newly electrified streetlights. When the
British imprison Denge for political subversion, Yasmin travels
to the mainland to fetch a mimeograph machine to produce a
Communist newspaper on Zanzibar. The tug of war of the title
is the one between oppressors and oppressed. Barracking from

the jubilant audience (they had come to life when the speeches ended) dissolved frequently into the night sky, especially when a new actor, evidently known to many, entered the frame. The line 'fishermen don't fall in love' drew thunderous laughter. As the film reached its elegiac ending, Yasmin's boat approaching island sands in moonlight, another Zanzibari moon was high, and the crowd again fell silent for a contemplative moment, before erupting into tumultuous applause.

In Forodhani Gardens at six on a Friday night boys jumped off the sea walls and rows of stallholders touted fish skewers, cats waiting below the trestles. Mostly, in the day, people sat around. Unemployment, I was told, runs at 65 per cent on Unguja, the Kiswahili name for the main island.[1] When it was cool I walked through layers of silted cultural debris left by tides of immigrants. Zanzibar has more than fifty mosques, an Anglican cathedral, a Catholic cathedral, a Hindu temple and the oldest Jain temple outside India. Bamboo scaffolding propped up Evangelical chapels jammed into gaps in the narrow lanes next to a slaver's mansion and shops owned by an embedded population of Indian descent. I had a skirt made in a New Town alley opposite Creek Road by a fifth-generation Gujarati who still spoke the tongue of his ancestral homeland. His forebears had brought the famous teak doors to Zanzibar, heavily worked with symbols identifying the resident. The door to my own apartment displayed a thick chain round the edges, denoting the island's links with slavery. Gujarati craftsmen incorporated bronze spikes into their doors to keep ramming elephants at bay, even though no pachyderm ever roamed Zanzibar's flatlands. And it still goes on. Every time I took the road south I saw carvers working mango-wood doors, some sprouting elephant spikes; even women sat carving. A new layer.

---

1. 'Zanzibar' is the English name of the whole archipelago as well as both the main island and its capital.

I had intended to hop on a *dala dala* public minibus north to Nungwi. But the Omicron apocalypse dominated my newsfeeds, and I quailed at two and a half hours in a packed vehicle in which nobody wore a mask. So I took a car owned and driven by the amiable Spiderman, an island fixer to many visitors before me. (He had adopted the name as a marketing ploy. His real name was Nyange.) I ended up travelling a good deal with the Spider, learning about the five children at home in Chuini who included one set of twins, the parents in Ng'ambo, the new town (the Other Side in Kiswahili, meaning the other side of the creek), and Spiderman's loyalty to Zanzibar. He had never been to the mainland, even though it was only twenty-five kilometres away by ferry. 'Why should I go?' he asked as we set off up the new Chinese-funded road. Exiguous *duka* shops broke up the banana trees and coconut palms, along with bursts of papaya, cassava, sweet potato, breadfruit and sugarcane. In one *duka* a row of tinned mandarins caught motes of dusty sunlight. So that was where they had gone. Memories wait patiently in surprising places. Women sold mangoes from small tables, or from the dirt at the side of road, and when it began to rain, their children deployed banana leaves as umbrellas. After we bought fruit, Spiderman picked a quinine leaf for me to try ('Bye bye malaria!'). Before the Kisiwa sugar refinery at Mahonda we passed more of the East German-built accommodation blocks I had seen in town, whole villages, really, which had risen in the heady days when the two governments lived in socialist brotherhood. When I asked Spiderman if these homes were popular with residents, he said they were, but that only those in town had running water.

In Nungwi shipwrights were caulking dhows with kapok and tempering planking. Alongside them, fishermen mended nets with wooden needles like giant pencils. For the indolent, a CHANGJ sign under a sheet strung between bamboo poles advertised a net-mender. Sheltered from the sun under the canvas roof of a café, men ate rice cooked with green bananas while others slept till the next shift. On the beach, coral-block anchors lay next to the dhows, each vessel stowed with a detached rigger, a baling bottle and a raincoat. A fish-gutter sat cross-legged on a stone slab with a bucket between his knees. The secondary identity of Nungwi unfolded at its western end. On the beach a Serbian tourist had swaddled a tuna in towels and lain it on a lounger. He showed me a film on his phone of him catching it. He had brought his mother on holiday and they were staying in one of the small arc of monster hotels extended round the bay. A number of Maasai strolled up and down the sand holding hands with white women. I had been forewarned of this phenomenon. Some of the Maasai looked like Maasai. Others resembled James Corden. The beads they wore round their necks, wrists and ankles were the cheap ones I had seen in shops. The Maasai wore nylon shorts under the red suka robe. 'Plastic Maasai,' a friend in Stone Town had told me dismissively, 'from Arusha. Go for *jigi jigi* with Italian women.' I referred to a version of this phenomenon in Chapter 8 after an attractive man in Bishkek market took my hand over a mint tea. Meanwhile, on Nungwi beach at three in the afternoon, a madrassa emptied, and little girls in white hijabs fanned over the sand.

On the way back Spiderman stopped the car at Kivunge hospital. Knots of female visitors waited in the wide corridors. A sequence of mops stood at angles on a balcony, positioned at the barre for a soap opera. On the second floor I looked into the window of a locked room packed with new electronic medical equipment sheathed in heavy plastic. Discarded polystyrene and cardboard wrappers lay jumbled on the floor. A UK-based medical charity manages Kivunge on behalf of the Zanzibari government. I hoped the equipment was not a symbol of the failed aid I had seen elsewhere in Africa. In Chapter 4 I referred to capital schemes that rise like Ozymandias, never finished, or finished and never used, or used and then abandoned when the machines stop working and there are no spare parts. According to the World Bank, in 2019 Tanzania received $2.1 billion net in Official Development Assistance (ODA) and other aid. *Look on my Works, ye Mighty, and despair!*

I had arrived in Tanzania the day after its sixtieth anniversary, 10 December 2021. The government celebrated with an announcement heralding new press freedom. The current president's predecessor, John Magufuli, was an enthusiastic suppressor of the press, shutting down or suspending critical newspapers and websites, and arresting journalists. Magufuli had taken office in 2015 as a representative of Chama Cha Mapinduzi (CCM), for decades the dominant party, on an anti-corruption ticket promising investment in industry. He supposedly lifted Tanzania out of the low-income category – according to the IMF, during his presidency Tanzania experienced 6 per cent annual growth. To achieve these goals, in 2016 Magufuli suspended the Swahili-language paper *Mawio* for alleged false reporting over the nullification of election results in Zanzibar. His re-election in October 2020, according to the *Journal of Democracy*, constituted 'an authoritarian landslide, achieved through electoral manipulation unprecedented in both scale and audacity. This was accompanied by high levels of violent oppression.' The regime also mobilised the army to Zanzibar that year ahead of the archipelago's own presidential polls, targeting opposition activists. On the second island in size, Pemba, security forces fired on demonstrators attempting to prevent the transport of allegedly fraudulent ballots, killing as many as nine. Human rights group Freedom House awards Tanzania thirty-four out of a hundred for its record on civil rights and political liberties. Magufuli was against family planning, inventing and promoting the slogan 'Set Your Ovaries Free'. He was also a Covid denier. Hospital doctors had to refer to 'respiratory problems' rather than the virus. According to *The Economist*, in March 2020 the president, a Christian, announced that, 'Corona is the devil and it cannot survive in the body of Jesus'. When Magufuli instructed security forces to test papaya and sheep with PCR kits, it turned out that all the fruits and ruminants did indeed carry Covid-19! After three days of national prayer, in June 2020 Magufuli declared his nation Covid-free. NGOs positioned themselves as 'promoting sanitation and disease prevention', as if they publicly stated a goal of Covid-mitigation, the government might kick them out. Then sixty-one-year-old Magufuli suddenly died, according to official sources of 'chronic atrial fibrillation'. I perceived generalised shock even nine months later. The president's death had certainly come as a shock to vice president Samia Suluhu Hassan, a Zanzibari native universally known to voters, since her abrupt elevation to the presidency, as Mama Samia. She reversed the country's position on Covid. The vaccine rollout has been agonisingly slow. Tanzania does not release Covid figures.

In late afternoons I walked away from the Swahili gothic of Stone Town lanes, along the shadeless beach fringed with plastic bottles, past another hospital and decaying government offices. A small but unfriendly warship looked on, and a sign, like one hanging outside a pub, warned in English against photographing or 'conducting activity' in the zone. I stood at the water's edge and stepped out of my sandals, looking towards the mainland. People said you could see mountain blocks of the Eastern Arc on a clear day – Nguru especially. But I never did. I was engaged in this pleasant reverie when a man's voice barked, 'Don't swim here.' He was standing in a Zodiac next to another man, and they had apparently, judging from the cracked trail in the water, motored over from the warship. Both carried weapons. 'I'm not swimming,' I said before I had time to reflect that this was not the moment for conversational engagement. So I was not actually invisible yet.

The flat main island of Unguja, eighty-five kilometres long and thirty-nine kilometres wide at the waist, is about the size of

Lesbos, Skye or Oahu. The population is nine hundred thousand, about the same as Oahu, eight times that of Lesbos and ninety times that of Skye. Facing proudly east, and benefiting from favourable trade winds, Zanzibar has enjoyed links with the rest of the world for millennia: in the tenth century, craftsmen in Siraf on the Persian Gulf raised houses using timber from 'the land of Zanj'. Arab, Persian and other Indian Ocean merchants traded goods from the African interior with Swahili people of the coast who flew the sultan's red flag from their dhows. Portuguese mariners had reported landfall as early as the sixteenth century, and representatives arrived swiftly from Lisbon, initially taking control of the archipelago in a loose alliance with tributary sultans. But Zanzibaris grew tired of the Portuguese after 150 years and asked for Omani assistance. That marked the beginning of the sultanate of Oman. Seyyid Said bin Sultan al-Busaidi (1791–1856), fifth of his dynasty to rule Muscat, capital of Oman, chose Zanzibar as his base on account of its proximity to Bagamoyo, terminus of the caravan route to Lake Tanganyika. Kiswahili is still widely spoken in Oman.

Slavery had been part of East African life for many generations, but when Arabs and Europeans got involved the trade increased a thousandfold, and by the middle of the nineteenth century captured men and women from Central Africa accounted for two-thirds of the Zanzibari population. Slavers and ivory traders were operating deep in the interior by then, and Zanzibar had grown to become the operational centre of both businesses: its central slave market was East Africa's largest. A forest of rigging had sprouted in the harbour where I watched girls swimming, and the sultan sat in his palace while men, women and children stood in chains, waiting to board triple-decker dhows bound for the Persian Gulf and Arabia. 'When you play the flute in Zanzibar,' the Arabic proverb went, 'all Africa dances.' A demand for billiard balls, piano keys, combs, knife handles and much else had stoked the ivory market in Europe, and traders in Africa found they could profitably combine tusk-harvesting with people-enslaving. A slave might carry a tusk to market, for example, and then be sold himself. Captives travelled yoked with forked sticks. Others carried supplies, and in addition many Nyamezi men worked as paid porters for the thousand-strong caravans, emptying their villages. One Nyamezi witness left a report:

> Every day we came upon the dead, and certainly we witnessed not less than a hundred deaths. Men were killed by the club, or the dagger, or I with my own eyes saw six men at different times choked to death: the victims were forced to sit leaning against a tree; a strip of bark or a thong was looped around the stem of the tree, pulled taut from behind, and the slave strangled.

The witness reported owners' servants clubbing children between the eyes, 'which bespattered their faces with blood', and throwing newborns into the bush. Slaves not shipped out from Zanzibar worked the clove plantations, where mortality was high – as high as 20 per cent. Tippu Tip (1837–1905), African Omani by descent and probably Zanzibar-born, specialised in the procurement of clove slaves. Tippu, who at one time controlled the whole region between Lake Tanganyika and the Upper Congo, was often in Zanzibar, working with a succession of its sultans. The handsome, mottle-walled house where he died still stands near the Airbnb I took in Shangani on the western bulge of Zanzibar main town.[1]

The British government regarded the East African coastal slave trade quite differently from the western one that despatched men and women to the cotton plantations of Georgia's Sea Islands described in Chapter 5. Britain's representatives seemed to think (if that is not too active a verb) that the country bore some

---

1. His actual name was Ḥamad ibn Muḥammad ibn Jumʿah ibn Rajab ibn Muḥammad ibn Saʿīd al Murjabī. Yay.

responsibility for west coast trade, having been heavily involved in it, but that east coast slave business was in Arab hands. This characteristically specious argument meant little to enslaved peoples on either coast, and anyway self-interest was at play, as it always was: Britain enjoyed friendly relations with the Sultan of Zanzibar which politicians were unwilling to jeopardise through the promotion of abolition, not least because any consequences might enable France and Germany to gain a foothold in the territory, already an obvious gateway for the looming Scramble. Notwithstanding this attitude, Britain signed treaties with the Sultan limiting the slave trade – a world-class case of weak-kneed compromise. In 1845, one agreement confined the movement of slaves to territory within the sultan's own East African possessions, and nineteen years later another prohibited the removal of slaves by water during certain months of the year. Navy vessels sailed self-righteously up and down the coast on the lookout for naughty slave captains. 'It became well known to the many officers on the east coast,' wrote Captain G. L. Sulivan, who spent five years in Her Majesty's service hunting slavers, 'that fine points in the treaties with the petty sultans and puny potentates . . . have acted as an impassable barrier against the abolition of the iniquitous trade.' In *Dhow Chasing in Zanzibar Waters* (1873), Sulivan records Arab skippers taunting British captains as they glided out of harbour, holds jammed with human cargo. He reckoned the annual number of incoming slaves in Zanzibar in the late 1860s to be as high as sixty thousand. 'Yet England,' he wrote, 'had become drowsy and unconcerned.'

This was the England I knew. Contemporary Tory politicians regularly trot out the dying John of Gaunt's speech in *Richard II*, the one about *This royal throne of kings, this sceptr'd isle,/This earth of majesty, this seat of Mars* . . . They omit the words with which Gaunt continues: *England . . . is now leas'd out . . . /Like to a tenement or pelting farm*, and . . . *bound in with shame,/With inky blots and rotten parchment bonds*. 1399, 1868, 2023 . . . when would it end?

Just like the slavers, white men who sailed in to 'open up' East and Central Africa in the last third of the nineteenth century depended on Zanzibaris for supplies and personnel. The Wangwana, Black freedmen of the archipelago, travelled widely on explorers' expeditions, mostly as carriers and *pagazi*. Henry Stanley said they were 'clever, honest, industrious, docile, enterprising, brave and moral'. Women marched with young children, and gave birth to more of them. Stanley once wrote that they had 'transformed stern camps in the depths of the wilds into something resembling a village'. But he brought only half of them back. As for supplies, the red-turbaned Banian, Indian traders and brokers who were mostly high-caste Hindus, sold cloth, beads, two-inch and one-inch screws to hold portable boats together, hinges and bolts for rudders and Dr Collis Browne's Chlorodyne stomach medicine. Dhow-chaser Sulivan recorded buying goats, tea and porter in a store 'like Thackeray's Mr Crump's establishment'. The whole community in Zanzibar was involved in the white man's business – African, Indian, Arab, European and American, the latter including the staff of a Boston trading firm. Engravings show explorers sitting anxiously at night on Stone Town roofs having collected mail at the Consulate. It was on a cool Zanzibari roof, in 1877, that Stanley learned his fiancée had married someone else (can she really have written, 'But his is here'?). It the only thing that makes one like him.

Zanzibari men were similarly involved in the slave trade. In the late 1860s the sultan himself derived £20,000 a year from human trafficking and maintained large trading posts in the interior. Even David Livingstone, a passionate abolitionist, could not afford to alienate the potentate: the pair sat sipping sherbet while a brass band played 'God Save the Queen'. The Blantyre man believed that Christianity must flourish in Africa, that to do so colonisation must thrive, and that in turn slavery must be abolished. Livingstone saw himself as more than an explorer. Perhaps, in their hearts, all 'explorers' do. Eventually, after the Atlantic slave trade had ended, Gladstone and parliament did decide that slavery in East Africa must stop too, and, after Britain threatened a naval blockade, it did. The Zanzibar market closed for ever in June 1873. Livingstone, who had died a martyr of a sort a week previously, would have been gratified to see an Anglican cathedral rise on the very spot where so many men and women had been sold like elephants' tusks. But it was not entirely over. The sea trade had ended, but domestic slavery persisted. In Zanzibar Indian merchants, the sultan and many other individuals still owned thousands of slaves. Hundreds worked on the clove plantations. An African trainee missionary in Zanzibar reported in 1893 that 'slaves are not sold publicly in the open streets as they were then, but in private houses. Several of these are to be found within a short distance of our own Mission House at Mkunazini.'

A new battle for control was under way. When the diminutive fabricator Stanley returned to Zanzibar in 1887, he was surprised to see German warships bumping alongside Royal Navy vessels in the harbour. Two years earlier, Germany had declared a protectorate over almost all the sultan's territory on the mainland, from the coast to the great lakes. The Scramble had begun, and in 1890 Zanzibar became a British protectorate under the sovereignty of the sultanate. For the next two generations Zanzibaris laboured under the stewardship of blimpish rulers like the one exemplified in *Vuta N'Kuvute* by Inspector Wright. The arrangement, and the protectorate, came to an end in 1963 when the island achieved Denge's longed-for independence. Revolution followed, as it so often does. Some twenty thousand Zanzibaris died – a phenomenal 10 per cent of the population, the equivalent of a million falling in battle in London today. Tanganyika had achieved its own independence from the British Crown in 1961 and become a republic in 1962. In the aftermath of its revolution, Zanzibar joined with Tanganyika to create the new nation of Tanzania.

The twentieth-century grail of nationhood, and a seat at the UN. But at what cost?

The *dala dalas* I never took disgorged alongside Darajani Market, as did deregulated *boda boda* buses, the whole traffic jam jostling against the bullock carts which trundled along every road. The market was the beating heart of Stone Town. Imported dates pressed stickily against the Perspex screens of wheeled stalls alongside exiguous tubs of black halva, and when I proffered a modest banknote (shillings were selling at 3,114 to the pound), the vendor wrapped a mound of dates in yesterday's *Nipashe*. However many I got, I ate them all at once. In the fish section, sellers squatted on the stone walls round three sides of their catch, flapping away flies. One man had arranged his blind-eyed fish in geometric patterns. The meat section was not for the faint-hearted. A morning downpour caught me out in Darajani. It was a Friday in jackfruit season. In an alley behind the tin-roofed section, women were waiting for the weekly charitable handout of pilaf rice. The downpour hammered so loudly against the iron roof that hundreds of people stopped talking. The rain beat all the noise in the world into the earth. Lanes ran like rivers. I ducked in and out of covered alleys, pressing myself against pitted coral ragstone alongside an assortment of Zanzibaris looking towards the sky. The rain emphasised the sharp angles of the market walls and the edges of the iron roof. It intensified everything. A secondary downpour slid in sheets off the edges of green awnings. Some people pressed on, eyes screwed tight. One man wore a tin motorbike helmet. The sky seemed in such a poor temper – but then the rain stopped. A shaft of sunlight fell between a break in the iron, illuminating the eye of a cow's head on a butcher's shelf. Another rebirth. The crowd split open, like the black Pemba vanilla pods stallholders were selling in ten-gram plastic packets. For a moment, even the blanket of humidity lifted a crack.

*

'I simply took the koran and tried to copy the letters on the shoulder blade of a camel.' After this plucky start on the writer's road, Princess Salama bint Said (1844–1924), Zanzibar's first published author, strove to introduce literacy among her female compatriots. I saw the actual shoulder bone, with its densely carved script, in a room crepitating with fans off a baked Stone Town lane. Born in Mtoni Palace, Salama, known as Salme, was the daughter of the first Omani sultan of Zanzibar and Jilfidan, a Circassian slave. In her early twenties she said goodbye to all that, eloped with a German merchant, married him in Aden in an Anglican church ('it was apparently sufficient for the pastor to hear me pronounce "yes" to everything') and boarded a frigate to Hamburg, where she began a new life as Frau Emily Ruete. *Memoirs of an Arabian Princess in Zanzibar* appeared in German in 1886 and in English two years later. In its salty pages the princess writes of the *heart* she felt to be lacking in German education; it was 'more than heart – the word is the Swahili *mayo*, intuition, sense of rightness and empathy'. Salme-Emily emerges as a pioneering commentator who, pointing out the distorting European perspective of published descriptions of Africa and Asia, foreshadows Edward Said's critique of Orientalism, the western reflex of cultural condescension towards peoples and societies of the East. 'Even in this century of railroads and rapid communication,' she wrote, 'so much ignorance still exists among European nations of the customs and institutions of their own immediate neighbours.' Her personal neighbours were self-righteous anti-slavers, but, as Salme observed, 'the gap between rich and poor is a form of slavery in the cold north'. Her analysis of western thought rejected the coherent West narrative, one that persisted to the Plato-to-NATO free-world construct put about in the Cold War.

Most of Salme's peers could not travel beyond their front door. I saw the covered bridges connecting Omani mansions to facilitate female passage. This was not 'only' a fact of distant female history and faraway patriarchal misogyny. It still exists, like Plato-to-NATO faked coherence. In some Taliban-controlled regions of Afghanistan in 2023, women cannot leave their homes. Progress? Where? Not only England was 'bound with shame'.

A friend I made in the close cool of Zanzibari afternoons, forty-six-year-old Mariam, called Princess Salme 'an inspiration, ever since I learned about her in school'. Mariam was divorced, like almost every woman I met in Zanzibar, and had three children. Her former husband, in possession of two new wives, did not contribute to the household. I spent many hours on a *baraza* with Mariam and her friend Patma. They initiated me in the intricacies of the kanga, the wrap women and men wear at home. Each one has a printed message. 'This was our way of delivering news before WhatsApp,' said Mariam authoritatively. 'If one wife gives her husband a kanga, the next wife thinks of another message.' When we got onto the subject of our aspirations, Patma said she wanted to see Livingstone's grave in Westminster Abbey (only his heart had stayed in Africa). I thought that was queer. When I revealed my own dreams, I got the impression they thought them mighty queer too. We talked of our lives. The more I told them of my so-called achievements, the more their confidence in me flagged. My confidence in myself flagged. I supposed that to be the point of cultural exchange. We got on to female rights in general. The women spoke in low tones about abortions and contraception, both illegal but widespread. They knew sympathetic pharmacists and doctors. With them, I saw Zanzibar through female eyes, just as I had seen the Bronx from the low vantage point of the stoops. A belief in the uselessness of men united the pair. I found I had much to contribute. Yet in a contradiction exemplifying the mystery of life itself, Mariam and Patma were keen to introduce me to the Swahili ritual of *singo*, the traditional massage-and-scrub wives perform for their men twice a week. One hot afternoon they marched me off to Patma's auntie's house and there warmed sugar, spices, flower petals and *oud* incense in a wooden tripod. An aroma of crushed cloves wafted through the coral-brick room. Talking throughout in the mixture of Kiswahili and English with which we daily battled, Mariam lifted my shirt and shoved the tripod up it 'to make you attractive', her mien indicating this to be an uphill task. 'If you're one of four wives, you have to *compete*.' Then she adopted the dog-and-lamp-post position, gesturing that I was to emulate her, and thrust the tripod up my skirt. I wondered, as smoke seeped through the fabric, whether my knickers were on fire. Patma, meanwhile, had mixed a *singo* scrub and was demonstrating massage techniques on the legs of compliant auntie. 'But when you massage your husband,' she warned, '*no cloves on the willy*.' All three women drew their faces close to mine in the aromatic gloom to ensure I had understood the diabolical consequences of this rookie error. I wrote the tip in my notebook.

Another female acquaintance, thirty-two-year-old Laila, did my toes at a spa in the new town. Like all spas everywhere, it played music designed to make you spend, presumably on the basis that you no longer care if you live or die. Laila had a seven-year-old son whom an aunt looked after on the outskirts of Dar, whence the family hailed. (Aunts were trending.) Her parents were dead and she had never married the boy's father, as there were 'problems of tribe and religion . . . He is Muslim, I am Christian. Muslims have too much wives. He four.' He occasionally gave her money. Across the street, a photographer with a professional umbrella shade was shooting a young woman in a hijab, an outboard motor by his side. 'The Zanzibari workforce is mainly made up of women,' said Laila as we waited for the polish to dry. 'You see all these women working in shops, and men sitting on the *baraza*?'

We had arranged, as a family, to meet for Christmas in Zanzibar. Reg had just completed his first term at Manchester University;

Wilf, twenty-four, was working for an international political lobbying firm; their father remained the North American derelict he had always been. I had carted both my children round the world many times since they were in nappies, as I have indicated, and as adults they had not rebelled by refusing to travel south of the Isle of Wight, as they might have. They had turned into vigorous travellers in their own right, and each had set off alone to South America, doing what I had done, though I hope not being as fucking stupid. I was so looking forward to their visit and had prepared all kinds of treats, trips and events. But when I wrote my name on my box of anti-malaria pills the morning the first was due to arrive (they came separately), knowing the container would shortly be sharing the kitchen shelf with three other boxes, I missed the solitude before it had even ended. I grieved, like Margaret for Goldengrove, for something that had to wither in order for fresh shoots to grow. And they did grow. Leaving four boxes of Doxycycline on the counter, we enjoyed our Christmas Day meal on the roof of Emerson Spice overlooking the other rooftops as the sun set. The night before we had been to mass at St Joseph's Catholic Cathedral, having heard the choir from the balcony of our Airbnb on the first floor of an Omani townhouse. A caretaking family occupied the ground floor, and we exchanged cordial greetings as they lay on the tiles cooking chapatis on a single gas burner. The father slept in the corner, his head in permanent Rembrandtian gloom. When it rained the central hall turned into a paddling pool even though it was covered, as a storm drain voided there. Washing lines threaded over the mezzanine between us like a cat's cradle, and a motorbike was stabled permanently in the hall next to a lavatory in which light filtered through banana leaves beyond an open window and fell on the porcelain bowl, creating an installation like a Swahili Magritte.

On Boxing Day a crowd gathered on the beach in front of Livingstone's to observe an antiquated ro-ro cargo ship, the LC *Kasa*, delivering a consignment of Mama's security vehicles: a royal visit was evidently imminent. A rainbow arced over the mainland that morning, but still no mountain appeared. Japanese jeeps and cars lumbered down a shallow ramp onto the sand – one inevitably got stuck and had to be towed. The final car had blacked-out windows. A couple of armed policemen dispersed the crowd. As for Mama: we passed her residence later on the way back to Stone Town from Makunduchi, near the unfinished Supreme Court. But we never saw her.

Mwanakwerekwe, the largest market in the archipelago, stretched along the eastern outskirts of town as we headed south, clothes displayed in unfathomable heaps alongside single shoes and minor mountain ranges of mangoes. A man bent fixing a twin tub in a remarkable state of dilapidation. Reg had bought sliders in a *duka* in one of the alleys radiating from Darajani, and one had split after a single day. But Spiderman knew where a Goan cobbler sat on a milking stool, and the man sewed the plastic in minutes. We settled on the coast in the far south. Grey waders and egrets had already left wandering hieroglyphics in the sand when the first thread of light appeared over Makunduchi beach. At six o'clock, men appeared from different directions and approached dhows beached in the mud. They rowed the motorless lateen outriggers out in single file, black against the rising sun. The napped surface of the water caught the first rays as the boats moved off towards the horizon, where they vanished. Dawn is a declaration of hope every day. But what is hope, in old age? *Sunrise*, Emily Dickinson wrote, *Hast thou a Flag for me?*

On the coast, people offered Reg and Wilf weed when they weren't with me or their dad. I developed a not disagreeable sense of objective obsolescence. Perhaps it was helpful for the observer to be out of the picture. One day we went to Paje to inspect a seaweed farm. At a beach restaurant, a local man, not a Maasai and not dressed as a Maasai, was having a late lunch with a mature Italian woman. Her blue eyeshadow had melted and settled into wrinkles around the sockets. The pair sat in silence, looking at their phones. The *Rough Guide* had told us we were dining at the best restaurant in Zanzibar, but it was terrible. 'She's even older than you,' Reg whispered as we finished our catch of the day, a nameless wrasse of the Indian Ocean. I felt instinctively that a parental speech on ill-conceived ageism was in order. Before I had composed it, Reg continued, 'and she's even fatter than you.' By the time I regained the will to live, he was far out to sea, and I could only see the tip of his snorkel charting a course over the reef.

At low tide, the beach was creased with seaweed, like a slept-on sheet. The tidal range on the east coast was huge. When it was low, children collected winkles in buckets and their mothers worked far out tying and untying seaweed plants from rope strung between stakes, a method known as off-bottom peg-and-line. According to a 2018 UN Food and Agriculture report, in a good year seaweed farming generates $8 million in Zanzibar. Farming is women's work, at least the peg-and-line variety. At the plant I visited in Paje, workers told me epiphytes were on the rise, likely as a result of a warmer ocean. Elsewhere on the island, NGOs have begun training Zanzibaris in deep-water techniques to grab more of the seaweed market in China, Japan and Korea. But most women can't swim.

In Kizimkazi in the southwest, president Mama's home village, fishers congregated in a clearing strewn with empty bottles, clipping outboard motors to a drying pole. Offshore, a sequence of small, uninhabited islands had tops like mushrooms. When we snorkelled over the coral forest, translucent turquoise fish glided in and out of bars of sunlight. The reef ran from two-metre-high upside-down buckets with noodly coral sprouting from the top to delicate reedy rushes swaying in the current. I had been reading about the reef fish, but only the wonderfully named axilspot hogfish and a shoal of Indian Ocean steephead parrotfish obeyed the identification books. The smaller species all moved together, changing direction swiftly, as if in panic. It was still Omicron season, and the piscatorial action was like the news in the papers. Along the beach the ruins of an Arab settlement disintegrated into a deep tidal creek, and the murmur of the waves, such as they were, and the low rustle of palms joined in a ceaseless dirge charged with the unresolved melancholy of all beaches.

Reg took a three-wheeled taxi to a Chinese-built hospital on the outskirts of Makunduchi to get an LFT, at that time a predeparture requirement. It was on the new Chinese-built road forking up the east coast, next to a Chinese-built school. His taxi driver waited. When Reg came down from the first floor, a commotion had broken out, and people were running to and fro, among them the taxi driver, who had mysteriously donned surgical gloves.

After the revolution Zanzibar was 'one big slum' according to some sources, and 'closed down' for five years. Many fled to Oman. Several older residents told me of the shambas, or three-acre plots, that the first post-revolution president, toothbrush-moustached socialist (and son of a Zanaki chief) Julius Nyerere (1922–99) ('the father of the nation') gave families on which to grow food. 'Without the shambas,' a Vuga man sitting on a *baraza* under a neem tree told me, 'we would have starved. I mean, more of us would have starved.' In recent years, people have sold or built on the shambas, 'which is not allowed'. The notion of entropic disintegration lingered in the minds of older people, often connected with the pre-eminence of western culture and the erosion of its indigenous equivalent. This was a story if not as old as the hills, at least as old as the Chinese-built highways slicing hills all along the Africa sector of that country's Belt and Road Initiative.

Vuga men talked about the school dropout rate. According to the World Bank, 31 per cent of Zanzibaris aged between fourteen and nineteen don't go to school. The archipelago has one state and two private universities but lacks investment in vocational training. State investment is scanty in general. President Hussein Mwinyi had entered Zanzibar's State House in October 2020 on a Blue Economy ticket, but anyone who knew anything mocked the new Ministry of Blue Economy and Fisheries, which amounted, according to one journalist, to 'rape and pillage of the sea'. A newspaper editor I met cited poor management, leases of uninhabited islands traditionally used by fishermen, and general lack of knowledge – 'no one has a clue what it [the Blue Economy] means'. 'We have felt the wind of change,' said another irony-minded old hack. 'All you need to do is read and write Kiswahili to be a member of the House of Representatives in Zanzibar [which has seventy-six members]. Voters want morning bread! You can buy votes, and once you're in power you can do anything – anything. That's what "power" means here.' He shifted along the *baraza* we shared into the shade, as direct sunlight had found an opening in the neem branches.

A couple of weeks later I went over to the mainland, and at the end of my trip there bought a plane ticket to take me from Dar back to Zanzibar. At the terminal gate I waited three hours, along with twenty or thirty other passengers. Air Tanzania staff had no explanation. They said, 'We will definitely leave at some point.' The hot night ticked on. I ate an enormous bag of cashews. Eventually we boarded, and, once we had strapped in, six tall passengers in business wear swanned down the aisle (there was no first-class cabin). Tanzania lands at eighty-seventh out of 180 on Transparency International's Corruption Perceptions Index (CPI), nestled between Suriname and Vietnam.

When I asked a journalist to describe the relationship between Zanzibar and the rest of Tanzania he hesitated and then said, 'bloody'. The archipelago remains dependent on the mainland, notably for energy security. Zanzibaris often say they are poorer than their mainland compatriots, 'because we have to import more', but according to Statista, per capita GDP is about the same at $1,100 (2020 figures). I was sitting with my interlocutor in Lukmaan, a sweltering, noisy restaurant with an open-air first floor. He had ordered baked bananas. An Indian family hovered next to our table, waiting for us to leave. 'And don't send Tony Blair again,' my companion said as his fruit arrived in a pool of canary-yellow sauce. Blair had appeared on the island a few months previously after the president (of Zanzibar, not of Tanzania) invited him and his eponymous Institute for Global Change to advise on the implementation of MKUZA, the policy plan for reducing poverty and growing the economy by 2025. Another friend expressed outrage that the Zanzibari flag, a relatively new invention, has the Tanzanian flag stuck in a corner. When he and I watched a football match together at Mao Tse Tung Stadium in Zanzibar new town, he complained, 'All the best players go to the mainland. Yet on the rare occasion when a Zanzibari player achieves anything at pan-African level, it is Tanzania who get the credit!' On the way back to Stone Town he pointed to a large bank blazing the words, BANK OF TANZANIA, ZANZIBAR BRANCH, apparently feeling that his whole island home had been reduced to branch status. Once he had the bit between his teeth, everything annoyed him. So what about possible Zanzibari independence, I asked. 'If you talk about independence,' he said, 'you get a bullet in the back of the neck.' Not literally, perhaps, but you didn't want to go round shouting about that topic.

Yet despite the 'bloody' relations, Zanzibar is disproportionately represented in Dar's unicameral National Assembly. Fifty of its 393 members are elected from Zanzibar constituencies and a further five selected from the Zanzibar House of Representatives – a remarkable number, given that Zanzibar has only 2.5 per cent of the population. The tension between the archipelago and the mainland is predicated on history, not economics. On the tenth anniversary of the union, the advent of colour television had revealed a relationship already fractured for decades. On 12 January 1974 Zanzibar's TVZ became the first station in Sub-Saharan Africa to transmit colour pictures. On the mainland, Nyerere said it was a luxury the nation couldn't afford (yet); his vision of African socialism insisted anyway on a return to cultural roots, and for years Tanzania was the largest country in the world without television – except in Zanzibar. As the decades unfolded, satellite dishes sprouted in wealthy areas of Dar notwithstanding a 600 per cent import tax. In the nineties, official resistance crumbled, but by the time mainland Tanzania got round to instituting a national broadcasting service there was no money to fund it.

Econometric analysis recognises varying degrees to which ODA sustains modest growth in Tanzania, but everyone agrees that aid has not significantly reduced poverty. Interlocutors willing to talk openly (if anonymously) were ambivalent about China, by many miles the main donor. 'We have had relations since the revolution,' said a journalist born in 1950. 'We are like sister and brother.' Tanzania plays a role in China's Maritime Silk Road project, itself part of the broader Belt and Road Initiative. Talks have stuttered for a decade over the construction of what would be East Africa's biggest port, sixty kilometres north of Dar at the place from which slaves sailed in chains. The project looks set to go ahead, with Omani support. 'China funds us in so many ways: hospitals, water, stadiums, streetlights, schools, the Zanzibar broadcasting centre, terminal 3 at Julius Nyerere,' the seasoned journalist added. But there was evidently much he could or would not say. 'Trying to find the extent and truth of both Russian and Chinese investment in Zanzibar, and in Tanzania overall, is tricky.'

*

My flight home was cancelled due to Covid-related staff short-
ages. There was no need to go home anyway. I decided instead to
take an unplanned trip to the mainland – initially to Ruaha, to
join Wilf. He was still a wildlife enthusiast, just as he had been
when he cried in Kenya about the anteater. He had arranged to
spend the second half of his office holiday on safari in Ruaha
National Park in southern Tanzania. I had waved goodbye in hor-
rifically unmasked crowds as he boarded the ferry to Dar. I knew
how he felt as he set off on his own. I had been in that position
so very many times. Was it ever worth it? I never stopped asking.

To be ready for my own early-morning flight from Dar to
Iringa, I took the last ferry from Zanzibar and stayed in an airport hotel next to the railway line. The hotel turned out to be a
not disagreeable relic of Tanzania's socialist past, though most
things worked, bar the lights in the hall. It was cavernous, lacking adornment of any kind with the exception of a ruched velvet
tangerine cushion, the likes of which I had not seen since 1973,
and a thread of tinsel on the fire extinguisher in the hall – I knew
that, as I had to grope my way along the corridor to regain my
room after a nasty supper and I put my hand on it. After checking
in early at one of Dar's two domestic terminals, I enquired as to
the whereabouts of an ATM. A security guy dropped his work
feeding luggage through a scanner and drove me to the other
terminal on his motorbike.

Coffee-coloured rivers uncoiled beneath the twelve-seater to
Iringa. Warthogs gathered at waterholes before green drained
off the land and shadows of small clouds lay on the earth. Then
we saw the mountains of the highlands, and herds flowing so
that you could not see individual animals. The plane made an
unscheduled stop at Selous safari camp. A white man in jungle
gear boarded. Shortly after we took off, he belched like one of
those volcanic lakes in Cameroon the sudden poisonous eructa-
tion of which killed hundreds. I found Wilf installed in a budget
camp near Tungamalenga. The 'dominant tribe' in both (the

phrase Innocent, Wilf's English-speaking safari guide, used) was Hehe. Tribal identification didn't seem to exist in Zanzibar. Ruaha is a Hehe word. We had a fine time in the park with wildlife maestro Innocent. A lioness lay guarding a buffalo foetus which she and absent colleagues had removed from its mother with the surgical precision of a box-cutter – the dead female, tail bitten off, lay gathering flies with a neat hole above her rump. Downstream, the Great Ruaha River where crocs and hippos were basking eventually became the Rufiji. A red agama lizard eyed us suspiciously, skittering crazily over a hot rock. Freya Stark said agamas have jaws as big as Americans.

On the way back to camp three young Maasai men and a child were leading a hundred mixed-colour dewlapped cattle and two bulls down a river valley. A Gogo couple sorting sweet potato leaves into a washing-up bowl invited us to join them, bringing plastic chairs out from their brick hut. A toothless granny and a young son in a torn Spurs top joined in. It was New Year's Eve, and later we walked into Tungamalenga, where under an awning men and a few women were ladling millet and maize hooch from a metal pail. We sat outside a shop. Music played from several places at once and giggling young women stopped at our table. It was hot.

After leaving Wilf, I took a road trip down to Lake Nyasa. For three nights I stayed on a hill farm in Kisolanza an hour from Iringa. I had a piece to write to a deadline, and I was glad to keep still. The ants next to my hut never kept still: they marched day and night in a heavily policed two-way corridor. Then rain fell in tumults and the ants vanished. Where had they gone, after such dogged industry? Did they have an end in sight? Continuing south, my driver called the T1 – the Tanzam highway – 'the road from Egyptee to South Africa' (Kiswahili speakers like to add an 'ee' to the end of English words). Everyone was talking about failed rains despite the recent shower, and from the T1 I could see withered crops. People stood in their shambas wielding digging hoes. At Mafinga, a lumber centre, women with translucent pails on their heads walked to the timber yards to sell pineapple quarters, and boys held grilled corncobs on pitchforks to open bus windows. Tawny pallets waited for the lorries famous for gluing up the T1. Onne, my driver, was a Christian Bena. He owned his car, having saved then taken a loan which he paid back by working for the Peace Corps, ferrying volunteers to their designated villages 'far, very all over far'. We had a riparian picnic in Kitulo National Park. A man with five dogs was hunting dikdiks. As we ate samosas and pineapple, Onne told me that he had experienced foreign food through his interaction with NGO workers. He liked cheese, which is not part of the Tanzanian diet, but when he took portions home to his wife and five children they recoiled in horror. 'What I don't like though,' he said as a metre-long monitor lizard shot past us into the sedge, 'is when you put together with something in the middle two pieces of bread.' Later he told me that *kenge*, the word for monitor lizard, is a term of abuse.

The motorcycle taxi ranks we had seen all the way down thickened at Mbeya, southern Tanzania's biggest city. Then the Rift Valley dropped away on our right, and at Tukuyu in the lush Rungwe highlands we entered banana plantations. At Ipinda the Tanzam forked, one side signposted MALAWI, the other ZAMBIA. A school uniform market ran alongside the traffic, displaying royal blue skirts, white shirts and black trousers along with towers of exercise books and racks of hanging backpacks. Tin roofs dotted the foothills of the Livingstone Mountains beyond, then came tea hills, bright sponges of green cut up into neat squares.

Princess Salme was the first female Zanzibari author; Tanzania's (and the island's) first Nobel laureate in Literature was 2021 honoree Abdulrazak Gurnah, born on Unguja in 1948. Zanzibar is a brooding, offstage presence in his novel *Paradise* (1994), but the action is set in this green and mysterious Tanzanian interior. In the story Gurnah conjures, in prose alternately lush as banana groves and tinder-dry as Dodoma in October, the shadow of colonialism in East Africa just before the First World War. 'There will be no more journeys now the European dogs are everywhere,' says Mohammed Abdalla, an uneasy éminence grise who travels to the heart of darkness with Yusuf, the youthful protagonist. Misgivings about iron-hearted Europeans poised to eat everything up foreshadow Inspector Wright, the blimp in *Vuta N'Kuvute* who boasts that the British 'built' Zanzibar. In the current climate, *Paradise* the movie, filmed in the Rungwe highlands, would dazzle at the Oscars and even, perhaps, launch a DIFF in the capital to line up alongside TIFF and ZIFF.

It was night when we arrived in Matema, the Livingstone peaks to the east matt black against a sky pierced with stars. The lights of a line of dhows lay across the horizon on Lake Nyasa, but the beach dipped, and when I entered the warm water, they went out.

Morning revealed the Matema Shore Beach Resort, my billet. 'Resort' was a misnomer. There were no other guests, and nothing to indicate resort-style activities. At 7.30 half a dozen staff in blue uniforms met for chapel over the upturned dhow outside reception. I wondered what they did the rest of the time. There was a menu, but food was best foraged: when a man appeared dangling a bucket of still-twitching little fish, or carrying a big one by its tail, I learned to follow him, and point to the fish in front of the cheery cook, who worked over an open fire in an open kitchen. My iPhone thought it was in Malawi, and remained loyal to Malawi time. In the morning a man without an iPhone pushed out a dhow or a woman washed clothes in a sudsy pail which she carried back up the sloping beach on her head in pale lemony light. In the afternoon, everyone went into the lake to lather themselves and their small children. I watched their foreshortened figures from the communal balcony outside my first-floor room. In the early evenings I took long walks along the sand in both directions, one towards the village, the other to a river outlet where hippos sometimes basked and in the day a dhow taxi ferried women with goods to sell in Matema. A sense of continuity persists on the shores of the great African lakes, something lost in the butter-eating north. Peter Matthiessen observed in the 'rank green and imminent rains' of Tanzania, 'the loving attention to the moment that is vanishing in East Africa, as it has vanished in the Western World.' And that was in the sixties. 'There rises from this hidden world,' Matthiessen went on, 'that stillness of the early morning before man was born.' At night lightning speared the lake, drawing the mountains out of their shadow, and rain fell on my bed.

Eel-shaped Lake Nyasa, also called Lake Malawi, is the fifth largest freshwater lake in the world by volume. (Livingstone called it Lake of Stars, as when he arrived at night he too saw fishermen's lights glittering on the pale black surface.) I paid a man to take me in a motorboat knackered even by Tanzanian standards to Ikombe, a village without roads where all the women are potters. As we approached, thirty children ran to meet the boat. The men were absent, presumably fishing or sleeping. Three pigs and ten piglets snuffled round two communal kilns. One woman fetched a cannonball of grey lakeshore clay, sat on the earth and in fifteen minutes made a bowl. She did not have a wheel – she used her hands. Mains electricity had never come to Ikombe. Ground-level solar panels powered basic phones and radios; music emanated from several huts. I had observed panels elsewhere, but people always said the same when I asked about them. 'They rust.' Humidity runs at 82 per cent.

Nobody in Matema or Ikombe parroted *mambo!* or *poa!* like they did in Zanzibar when they saw a white skin. They said *karibou*. While *mambo* (hello, how are you?) and *poa* (all right, good) are perfectly normal Kiswahili words, the tourism industry in Zanzibar, which pre-Covid represented 80 per cent of archipelagic income, had persuaded everyone to sing them out whenever a visitor hove into view. Farouque Abdella, a London-trained

Zanzibar-born fashion designer, held court every day between
11.30 a.m. and 2 p.m. in a Stone Town hotel. Abdella remembered the elegant language of his youth. 'I grew up with classic
Kiswahili, none of this *mambo – poa*. In the old days it was *habari
za asubuhi* for "How are you?", almost Japanese in elegance. When
I left here in 1966, Kiswahili was like poetry – like Shakespeare –
now I have to ask what does this or that mean?' Barefoot little
boys came in for coins as we spoke; Abdella allowed it on Fridays,
and tried to teach them to say thank you.

For my last week in Stone Town I rented the first floor of a home
built in 1870 for a Goan merchant. The ceilings were three
metres high, and in the central courtyard, now iron-roofed,
they reached thirty metres. The smooth stone floors, cracked
in places, appeared matt and marbled. At some stage a resident
had jemmied in an air con unit which didn't work and installed
a wall-mounted television (I don't know if that worked as I didn't
try it), but otherwise the place was unmodernised and furnished
with dark, hardwood pieces scalloped and carved in the Swahili
style. Teak clocks which had long since lost interest in bour-
geois banalities of time hung their hands limp at 6.30. Niches
of Moorish design contained a solitary glazed pot or a metal
tray a metre in circumference, and tall, waisted clay vases stood
haughtily in corners, filled with elegant branches that reached my
chest. No mug rings stained the tables, and no modern detritus
chased away Goan spirits. The seat of a chair at the foot of the
four-poster had been rewoven with fishing thread. The front
windows looked across an alley into the abandoned ministry of
education (they had built a new one near the airport), a few panes
glinting in the fanlights and bleached wooden shutters flapping
in a monsoon wind. Next door to that, a mosque broadcast the
call to prayer down the alley and those radiating off it, all busy,
except during the midday heat. I set the battered postcards in
niches and nooks. Even as I chose the right spot for Botticelli's

Medusa-haired Venus, her expression no less wistful as the decades (and centuries) passed, I knew, because I had learned it since the igloo, that happiness comes only in the present tense. But the ritual was too old to be abandoned.

Layers of history had leaked and seeped and Zanzibar asserted itself as a unified entity. Until the end I caught sight of the ghost of John Sinclair in the streets. He was chief secretary to the protectorate government and, as a sketchily trained architect, had a hand in many colonial buildings – the Post Office, for example. Sinclair lived in Zanzibar from 1899 to 1923, and he had a little Raj in mind. Even on my last day I stumbled on one of his private commissions, the Bharmal merchant's house near Darajani in which he combined rounded and pointed arches. Faded bank notes signed by Sinclair come up at auction sometimes, but his plan to Europeanise Stone Town failed. Zanzibar took over. The architect's hand remained, but Inspector Wright and his goons from *Vuta N'Kuvute* really had vanished like dhows on the horizon. Meanwhile, round the corner from the Bharmal, the deco Cine Majestic was screening Africa Cup matches. They had chalked the fixtures on a board outside. The Majestic had functioned as an open-air venue since the roof fell in twenty years ago. It had layers of its own. John Sinclair, of course, designed its predecessor, the Royal Cinema Theatre. Fire destroyed that in 1954, just after my landlord's father had taken his future wife there on their first date.

Tanzania had banned plastic bags, and mangoes came in small biodegradable sacks. But single-use plastic water bottles littered the beaches. This exemplified a more generally confused picture. From a low start at the first COP (Conference of the Parties) in Berlin in 1995, the situation has deteriorated. Since that call to arms, we have released into the atmosphere half of all the carbon produced since Egeria rode her donkey into Jerusalem. But one thing has improved: public awareness. Back in 1995,

The New York Times reported that 'only a few' scientists attributed warming (then up 0.6°C since the Industrial Revolution) to
human activity. Now everyone bar loons accepts that we stand
on the threshold of catastrophe due to anthropogenic carry-on.
A normalisation of climate alarmism means most people believe
hundreds of millions are set to become climate refugees; that
drought-stricken Madagascar faces the first climate famine; and
that ice sheets may have reached a tipping point. It helped when
the problem encroached on rich nations: the Pacific Heat Dome
of 2021 killed a few thousand in the US, breaking temperature

records by devastating margins. True denial is in final retreat, even as fresh decarbonisation targets whizz by unmet and the most recent COP, in Glasgow, became known in some parts as FLOP26. The world had caught up with Rachel Carson. At about the time Tanganyika and Zanzibar joined forces she had written, 'man is part of nature, and his war against nature is inevitably a war against himself ... I think we're challenged as mankind has never been challenged before to prove our maturity and our mastery, not of nature, but of ourselves.' I wrote in Chapter 5 that cheap energy is a thing of the past. It is still far from certain that we have listened to Carson as we seek to solve our dire energy problems.

Environmentalism had entered the educational debate over what children should read, a war in which words are bullets. In 1971 Dr Seuss, whose real name was Theodor Geisel, published *The Lorax* – a *Silent Spring* for under-tens. 'I speak for the trees,' the Lorax says as he defends a Truffula forest scheduled for the chop. In 2021 the fiftieth anniversary of the book passed almost unnoticed, according to the *New Yorker* 'Dr Seuss himself having been made into something of a thneed in the latest round of book battles'. (A thneed, in *The Lorax*, is a 'Fine Something-That-All-People-Need' and all the Truffula trees topple to make it.) After a child from a North Californian logging family brought *The Lorax* home from school, a community group paid for an ad in a local newspaper calling for the book to be withdrawn because 'to teach our children that harvesting redwood trees is bad is not the education we need'.

The electricity was out at Livingstone's one morning and the owner, Glory, had set a tripod on the metal counter. Under it he made a small charcoal fire over which to stand a copper kettle. The beach was stirring to life and the rising sun far away, lofting indifferently over the Indian Ocean. A man I knew was waking up on the sand. I had just been standing under a blossoming flamboya tree outside the Serena where I regularly picked up the hotel Wi-Fi to read the *Guardian* on my phone. The top story revealed that homelessness in the UK had doubled among nationals in a decade, and in addition, we were not able to house refugees in acute need. War in Ukraine was two months ahead, but news that morning concerned the thousands of unhoused refugees in the UK, many with right-to-remain status. A percentage were living rough. I had been hosting refugees for several years through a Glasgow-based NGO, and one of my guests had been living on the streets before he came into my house. He was from a large African nation, and told me how in London other homeless people stole his shoes. But he never really said what it was like to sleep on the pavement. The hosting experience had elevated me. I had seen my own infantile anxieties for what they were. I had seen too what little progress had been made. Why, sixty years on, did many in Britain express the opinions Tom Stoppard had recorded in the *Western Daily Press* in 1961? Why could people not see, as I had seen in the Bronx and Zanzibar and so many other places over all these decades, the ways in which immigration enhances and enriches a culture? Why had there been not progress but regression in a country that was once the champion of the world? As we saw in the Introduction, Stoppard had written, 'It is never the landlord who doesn't like coloured people. It's the other people in the house, or the other people in the street. No one you meet is guilty.' Now, half the country was willing to speak against immigration openly and on the record, and some had castigated the RNLI for picking up drowning migrants. A cartoon depicted a lifeboatman asking a person floundering in the water, 'What country are you from?' and the man yelling back, 'Earth.' The charcoal jenga collapsed under the tripod, and Glory poured the kettle onto a spoon of coffee powder. It is the writer's duty to have hope, but in my own land, one supposedly with glory of its own, it was hard. As Gaunt said, we were 'bound in with shame'.

Women meanwhile had regrettably broken no more

ecclesiastical barriers. The Vatican held out. But my mother continued to express carefully weighed views on other topics. Talking of her cousin, a woman whose offspring Sue is about my age, Mother said, 'That Sue: she's a daughter in a million.' When she first said it, the suction pump of my subconscious did its work and I forgot it. But when she said it a second time a year or two later, it stuck, I suppose for life. Like many, Mum was a natural surfer on the populist wave. To stoke a multitudinous cohort, news outlets deliberately fostered a belief that either the elite or the left was ganging up against ordinary people as well as letting in migrants with whom they did not share the bond of kinship. To a certain extent environmentalism and climate action had provoked the reaction, just as it had goaded those militating against the defenceless Lorax. The colossal deficit spending of the pandemic had not just bailed out bankers in the 2008 emergency payout style; it bailed out society as a whole, but that turn of events failed to dampen populist passion. Anti-vaxxers had recently gained traction across Europe and North America. Since I'd left the Bronx, populism had both weaponised racial divisions and unleashed the Capitol attack. It was impossible for an old person like me, looking at my children, to imagine that their generation would ever be able to normalise the debate. Real drivers of change had not appeared under boomer stewardship. Economic inequality had if anything got worse. I saw this at a granular level in my involvement in the well-being of my brother Mathew. In his forties his behaviour had deteriorated. His neighbours in Weston-super-Mare, where he lived in what passes for him as relative calm for ten years, obtained an Environmental Protection Order and got him evicted. By this stage very few care facilities were equipped to handle his violence. But we escaped the secure unit five hundred miles from us and everything he knew, just. I made the fourteen-hour round trip each month to his new home in South Wales with love and a heavy heart. We no longer look alike. But when we sat the other day between a standard lamp and a white wall, I noticed that our shadows are like twins. I had seen, through interactions with Matt's cohort, that as the vulnerable were marginalised, they had become more vulnerable, setting up a cycle of degradation and a society in which winner takes all. The deeper we moved into the neoliberal age into which I was born, the worse things got. 'Greed is good' had arrived before I came of age, and unlike other era-slogans ('Fuck the Man' in the sixties), it had not gone away. The joy I experienced in the Shanghai hotel room in May '97 seemed part of the ancient past of the world, not just of my own short past. Over the two and a half decades that followed that Chinese dawn, ideas of freedom and liberal democracy – the cornerstone of my education – had ceded ground. Back then the overwhelming majority of Europeans thought their children would have a better life than they had had. Now most think the opposite. Market forces had won out over morality and boats had not risen on Thatcher and Reagan's tide of affluence. In the second of the decades, £30 billion has been cut from unemployment benefits, rent subsidies and social welfare payments in Britain. The average family income has fallen by 6 per cent, while inflation has climbed, first steadily, then precipitously. The gap between the life expectancy of the rich and the poor has grown to eighteen years. These inequalities stoked populist fury. It had looked as if the dawn of my middle years would end in crepuscular retirement. But I sense ethical midnight.

In my last week in Zanzibar a 1,600-word book review fell due. The material refused to submit, so I decided to look at it in hard copy. (I find seeing words on an actual page can solve some of the problems of prose. And you can only ever hope to solve some.) The Serena Hotel on Kelele Square had a business centre, besides offering free-in-the-street Wi-Fi. Two smartly uniformed young women on reception looked at me when I enquired about printing. One said, 'A dollar a page.' This was an extortionate rate even in a good hotel. So I said, 'How about fifty cents?' They immediately said yes, this time in unison, before a crisis unfolded involving a spilled bottle of nail polish. As I had only five pages to print I had no intention of arguing further, and I took my memory stick into the business centre and printed the recalcitrant review. Moritz Thomsen (1915–91), cited at the head of this chapter, wrote in *The Saddest Pleasure* (1990) that 'True communication between cultures is a chancy thing especially where money is involved.' The book tells the story of a journey taken 'raw and bleeding with loneliness' though Brazil. The 'chancy thing' lies at the heart of the travelling experience: toggling situations (or ideas or economic statuses) not comparable in any context. (I said in Chapter 6 that returning to a place involves a similar toggling manoeuvre, between the altered vista and the altered you.) The dollars at the Serena presumably went straight into the young women's pocketbooks so could I really have baulked at giving them five? However many decades you put in on the open road, you couldn't learn a strategy to deal with this issue. It is just like that, and all you can do is stand back, walk round it, keep working at it. And remember always to look through the eyes of the other, to walk a mile in their shoes, to deploy every analogy there might ever be, in order to be true, and not to live what Thomsen called his own 'life of lies'. The Seattle-raised author grew up privileged but repudiated his class and its money – the American dream in reverse. His moral stance had not brought spiritual rewards, nor any other kind. In *The Saddest Pleasure* he refers to Faulkner's 1950 Nobel acceptance speech, the one I cited in Chapter 5, 'I believe that man will not merely endure, he will prevail', and refutes it, believing that greed, racism and general human depravity give the lie to its optimism. So you think, *Whom do I believe?* Then you look at *The Sound and the Fury* and understand that Faulkner saw the horrors.

Thomsen wrote of the deep shade of mango trees and piled rocks on a shore, things I saw every day in Zanzibar. I was the same age he was when he wrote *The Saddest Pleasure*, a time when you know that most of the places you visit you will never visit again, and most of the books on your shelves you will never re-read. The knowledge lent every fresh scene elegiac melancholy in Thomsen's journey, and in mine. But there is something else. I had just read the book for the third time, noting with surprise the author's judgement of strangers, especially women – in Natal bus station he sees two young American women and comments on 'the grossness of their asses', and at Rio airport he calls some women 'whores' several times even though they are no such thing. It didn't make me like the book less. I liked it more, if anything, because the narrator revealed himself to be as flawed and failed and cruel as we know the world to be – as Thomsen himself had learned it was when his business partner kicked him off his farm in Rioverde, Ecuador, and out of 'that watery, half-drowned world that held the secret, the mystery of man's essential inconsequentiality'. He had wreathed a tree outside his window on the Rio Esmeraldas with spent typewriter ribbons. I still thought he was a terrific writer. The book is starred with poetry (the phrase belongs to Virginia Woolf). The landscapes of the *sertão* leap off the pages and the mournful echoing cadences of the style remind you of a great badly lit railway station where people are saying goodbye. When I went to northeast Brazil, Thomsen came with me.

I reappeared at the Serena reception desk to pay the $2.50 due and one of the women said, 'Oh, don't worry about that! *Hakuna matata!*'

In summing up changes in the travelling life, I allow myself the observation that ready access to news has altered the nature of absence. How often as a young woman did I sit on a warm rock twiddling the dial of a short-wave radio which may or may not leap to life with a crackle of a voice that made my heart beat faster? Or wait my turn at a *poste restante* queue? I can still feel the burning pain when a kindly postal worker in the Santiago main office said, 'Oh, I've sent many letters back for you,' as I had been away from the capital for too many months. I can still *see his face*. Place names themselves have gone: you really feel old when you remember somewhere as somewhere else. I went to St Helena on the *last mailship*. And coming home has changed: abroad was more foreign when I set out, as my Bristolian childhood indicated. So perhaps my supercilious reference to bullfighting posters in Chapter 2 was misplaced, and I should celebrate the unglobalised sixties. Home no longer comes as a shock either. Even in the early nineties when I returned from six months in Chile, I was annoyed, waiting for the 24 bus in Tottenham Court Road, when it didn't stop as I flagged it down, and I remember thinking, *Wait a minute, Sara. There are bus stops here*. Some things never change though, and my boys will endure them as I had to: the clenched jaw walking through Arrivals, where none of the signs or smiles is for me. And nostalgia anyway has always been with us. Fifty years ago Martha Gellhorn lamented, in *Travels with Myself and Another*,

> Remember when we were given generous portions of butter and jam for breakfast, not those little cellophane and cardboard containers; remember when the weather was reliable; remember when you didn't have to plan your trip like a military operation ... remember when you were a person not a sheep, herded in airports, railway stations, ski-lifts, movies, museums, restaurants, among your fellow sheep ... remember when you confidently expected everything to go well instead of thinking it a miracle if everything didn't go wrong?

Through it all, with some temporary betrayals, I remain loyal to the travel writing form. It is the perfect vehicle in which to smuggle the ineffable, the unsayable and the randomly comic. I mentioned Jonathan Raban's comments on the genre. After acknowledging that the most ancient of metaphors insists life is a journey, Raban wrote that he liked travel writing for the way 'it freely mixes narrative with discursive writing' and that 'the literary journey is more likely to be about time than place' – my sentiments entirely. Famously, Raban concluded, 'As a literary form, travel writing is a notoriously raffish open house where very different genres are likely to end up in the same bed.' I mentioned this in Chapter 8 and I agree with Raban, though I prefer 'raffishly notorious'.

As we all know, a changing climate, coronaviruses and the perils of voice appropriation have endangered travel writing and its ability to thrive. The form might even be going extinct. If I were starting out now, I would not be able to make the same choices. Young women resembling the one I once was are having to make decisions about their writing lives in the light of the world we are leaving them with, and I consider myself lucky to have set out when it was morally possible to take both the high road and the low road. I am confident those young women will find their own route, as I did.

The red-and-black notebooks, by the way, had disappeared for ever, Covid marking the watershed in this and so many other matters. I was able to accept the things I couldn't change and purchase any old notebook (providing of course it was hardcover A5, feint, narrow lined, with a margin).

In Zanzibar I realised all over again that if the observer wants to observe, she has to be alone on the road. As for the invisibility part of being an old woman, I couldn't care less. I will care more when I begin the next volume, following this Nubility to Invisibility trajectory: its title, as I indicated in the introduction, is to be *Immobility*. While I was in Zanzibar bell hooks died. She might well have used the mantra 'Set Your Ovaries Free', but in another context. I remembered with a rush, as I read the obituaries while sipping jammy Nescafé at Livingstone's, the influence she had on me as a young woman. 'To truly be free,' she wrote, 'we must choose beyond simply surviving adversity, we must dare to create lives of sustained optimal wellbeing and joy.' It seemed so simple at twenty. It had also seemed important to dare; now, I wondered if I had created the kind of life hooks had in mind (she took the name of her maternal great-grandmother, and instituted the lower case to distinguish herself from the elder Bell). Personally, I had much to regret; I like to think, having skin in the game, that everyone does. But it was not too late to seek 'sustained optimal wellbeing and joy' in old age.

In Chapter 1 I quoted Isabel Savory, who wrote in despair in 1900, 'I go to the ends of the earth; and behold, my skeleton steps out of its cupboard and confronts me there.' Yes indeed, there I was in Zanzibar. Savory also wrote that it was easy to tire of a new place, and that 'familiarity and close inspection betray the copper through the silver plate'. I had thought that to be a fine analogy. Now I see it really is, but not for the reason Savory intended. Copper's tarnished beauty is truer than the manufactured kind of silver plate, as well as more valuable, and in my forthcoming and final volume of travel memoir I will select authenticity every time. (Actually, I might call it instead *Burning the Candle at Neither End*.)

I also said at the beginning that I caught the end of second-wave feminism in the early eighties. I was twenty. We were still ecstatically absorbing the seventies burst of feminist energy. But the political strand of the movement was already weakening, and I said in Chapter 2 that I have seen feminist empowerment defanged in the West. Choice began to matter, rather than structural change. I never really saw feminism offering transformational change again, after the hooks years. In my adult life young women's idea of feminism has moved away from solidarity towards the agency of the lone actor. I teach at a college two days a week and when I told one of my Generation Z students that I used to go to a consciousness-raising group on the Cowley Road when I was an undergraduate at the same place as her, she smiled benignly as if inspecting a picture of a crinoline. No such thing as society indeed. I talked of seventies energy: remember *Our Bodies, Ourselves*? A hugely influential volume that made us believe, genuinely, that feminism could cross barriers of class, income, race and creed, promoting 'a sense of identity with all women in the experience of being female'. 'Lady, love your cunt,' Germaine Greer wrote the same year in *The Female Eunuch*. How did we travel to this land of depoliticised, even corporate feminism? In 2023 it can't even cross the barrier of sexuality. Second-wave feminism arose from the idea that the personal is political. That idea linked personal experience with broader societal and political structures. The intersectionality touted in the 2020s as a new thing is the same thing we talked about with excitement forty years ago. hooks was an apostle of intersectionality, but there is little evidence that my students or the wider world have absorbed its essential message: *Lean Out*.

I have come to the conclusion that the writer's duty is to strive towards hope rather than have it. Joan Didion had also just died. For obvious reasons, I had her advice about finishing a book in my mind. For writers, she said, it is 'sort of like Vietnam itself – why don't we just say we've won and leave.' I spent the last two days up at Matemwe, where my bungalow nestled in a band of gathering green above a cave that filled up when the tide was in, making the porous coral rock sing and sigh in the strangest way, just as Karen Blixen had described at Takaungu, not far up the coast in Kenya. A row of dhows ran noiselessly, though they no longer flew the red flag of the sultan. Far out, waves broke on the reef like a curl of lemon pith. In the bleached silence of midday I thought I might never go home again. Then at night the moon rose when it was already dark. It lifted as an orange ball, quickly becoming pale clementine as it crested the sky, sending a corridor of bright white light onto the ocean.

If life has any meaning at all toward the end, it is certainly only in one's ability to draw from the well of memory a satisfying amount of intense experience to contemplate, for the things that are forgotten are only symbols for the parts of us that have already died.

MORITZ THOMSEN, *The Saddest Pleasure*

Notes

Epigraph

ix Letter to Hortense Flexner, autumn 1942, cited Moorehead, Caroline, *Martha Gellhorn: A Life*, Chatto & Windus, London, 2003

Introduction

1 Murphy, Dervla, *Cameroon with Egbert*, John Murray, London, 1989
3 Gellhorn, Martha, letter to Alvah Bessie, 1958, cited Moorehead, op. cit.
3 Gellhorn, Martha, letter to *Collier's*, summer 1936, cited Moorehead, op. cit.
8 Stoppard, Tom, cited Kynaston, David, *Modernity Britain: A Shake of the Dice 1959-62*, Bloomsbury, London, 2014
11–12 de Beauvoir, Simone, *The Second Sex,* first pub: Gallimard, Paris, 1949, trans. H. M. Parshley, Alfred A. Knopf, New York, 1953
12 Friedan, Betty, *The Feminine Mystique*, W. W. Norton, New York, 1963
14 Bennett, Alan, cited Kynaston, op. cit.

1 At Least 300

19–21 Wheeler, Sara, *Terra Incognita*, Jonathan Cape, London, 1996
21 Scott, Robert Falcon, 'Terra Nova journal', 1911. Accessed: https://www.spri.cam.ac.uk/museum/diaries/scottslastexpedition/about/
21 Scott, Robert Falcon, letter to Kathleen Scott, 1911
22 Huntford, Roland, *Shackleton*, Hodder & Stoughton, London, 1985
24 Woolf, Virginia, *A Room of One's Own*, The Hogarth Press, London, 1929
25–26 Parrish, Maud, *Nine Pounds of Luggage*, J. B. Lippincott & Co., Philadelphia, 1939
26–27 *Peregrinatio*: the core of the text is preserved in the eleventh-century *Codex Aretinus*, which was rediscovered in a monastic library in Arezzo in 1884
29 Justice, Elizabeth, *A Voyage to Russia*, G. Smith, London, 1739
29–30 Wheeler, Sara, *O My America!,* Jonathan Cape, London, 2013
30 Dixie, Florence, *Across Patagonia*, Bentley, London, 1880

30   Horace, *The Epistles*, Book I, Epistle XI
30–31 Savory, Isabel, *A Sportswoman in India: Personal Adventures and Experiences of Travel in Known and Unknown India*, Hutchinson & Co., London, 1900
32   Brennan, Maeve, 'The Springs of Affection', in *The Springs of Affection*, Houghton Mifflin & Co., Boston, 1997
32   Bourke, Angela, *Maeve Brennan: Homesick at the* New Yorker, Jonathan Cape, London, 2004
33   Friedan, op. cit.
34   Letter of application to join Ernest Shackleton's *Endurance* expedition, 1914
35–36 Steger, Will, *Crossing Antarctica*, Alfred A. Knopf, New York, 1992
36   Parfit, Michael, *National Geographic*, Apr 1993
37   Peden, Irene C., cited Rothblum et al, *Women in the Antarctic*, Harrington Park Press, New York, 1998
37   Cited Sebba, Anne, letter to the editor, *Spectator*, 30 Jun 2007
37   Blum, Arlene, *Annapurna: A Woman's Place*, Sierra Club Books, San Francisco, 1980
38   Peden, op. cit.

2 MONI SOU?

47   al-Shaykh, Hanan, 'Cairo is a Grey Jungle', in Govier, Katherine (ed.), *Without a Guide: Contemporary Women's Travel Adventures*, Macfarlane Walter & Ross, Toronto, 1994
48–49 Donnelly, Julie, *The Windhorse*, Jonathan Cape, London, 1986
52   Cather, Willa, *My Antonia*, Houghton Mifflin & Co., Boston, 1918
54   Wheeler, Sara, *An Island Apart*, Little, Brown, London, 1992
57   Naipaul, V. S., *An Area of Darkness*, Andre Deutsch, London, 1964
58   Theroux, Paul, *The Stranger at the Palazzo d'Oro*, Houghton Mifflin & Co., Boston, 2004
58   Chesterton, G. K., 'The Advantages of Having One Leg', collected in *Tremendous Trifles*, Methuen & Co., London, 1909
59   Adichie, Chimamanda Ngozi, *Americanah*, Alfred A. Knopf, New York, 2003
59–60 Gellhorn, Martha, *Travels with Myself and Another*, Allen Lane, London, 1978
71–2 Woolf, op. cit.
72–3 Bruce, Mary, cited Stoneman, Richard (ed.), *A Literary Companion to Greece*, Penguin Books, Harmondsworth, 1984
73   Woolf, Virginia, *Three Guineas*, The Hogarth Press, London, 1938
73–4 West, Rebecca, *Black Lamb and Grey Falcon* (two vols), The Viking Press, New York, 1941
75   Wheeler, Sara, *Evia: Travels on an Undiscovered Greek Island*, I. B. Tauris, London, 2007 (revised edition of *An Island Apart*, op. cit.)

3 LADYLAND

78   Hitchens, Christopher, *The Missionary Position*, Verso, London, 1995
78   Hitchens, Christopher, 'The fanatic, fraudulent Mother Teresa', slate.com, Oct 2003
86   Lewis, Norman, *A Goddess in the Stones: Travels in India*, Jonathan Cape, London, 1991

87 Lewis, Norman, 'Genocide in Brazil', *The Sunday Times Magazine*, 23 Feb 1969

87 Naipaul, op. cit.

88–9 Gupta, Malvika and Padel, Felix, 'The Travesties of India's Tribal Boarding Schools', in *Sapiens*, 16 Nov 2020

90–1 Butalia, Urvashi and Menon, Ritu (eds), *In Other Words: New Writing by Indian Women*, The Women's Press, London, 1993

91 Review of *In Other Words* in Bardolph, Jacqueline (series ed.), *Telling Stories: Postcolonial Short Fiction in English*, Brill, Leiden, 2001

91 Chughtai, Ismat, *Lifting the Veil*, Penguin India, Delhi, 2009

94 Major, Norma, *Joan Sutherland: The Authorised Biography*, Macdonald, London, 1987

94 Major, Norma, *Chequers: The Prime Minister's Country House and its History*, HarperCollins, London, 1996

94 Butler, Judith, *Gender Trouble: Feminism and the Subversion of Identity*, Routledge, London, 1990

95 Birkett, Dea and Wheeler, Sara (eds), *Amazonian: The Penguin Book of Women's New Travel Writing*, Penguin Books, Harmondsworth, 1998

101 Gellhorn, Martha, *The Face of War*, Rupert Hart-Davis, London, 1959

102 Thubron, Colin, 'Sophisticated Traveler', a review of *Passionate Nomad: The Life of Freya Stark* by Jane Fletcher Geniesse, *New York Times*, 10 Oct 1999

103 Shan Price, Alison, 'Stepping Over the Horizon' (talk given on 12 Oct 2017), quoted in Goswamy, B. N., 'Freya Stark: the poet of travel', *Tribune India*, 16 Feb 2020

4 THE STORMING ARMY

105 Angelou, Maya, 'Our Grandmothers' from *I Shall Not Be Moved*, Random House, New York, 1990

105 Cherry-Garrard, Apsley, *The Worst Journey in the World*, Constable and Co., London, 1922

111 Blixen, Karen, *Out of Africa*, G. P. Putman, New York, 1937

112–13 Wheeler, Sara, 'A Traveller's Tale: And the Boys Came Too', *Guardian*, 10 Oct 2009

114 Murphy, Dervla, *Full Tilt: Ireland to India with a Bicycle,* John Murray, London, 1965

114 Murphy, Dervla, *In Ethiopia with a Mule*, John Murray, London, 1968

114 Murphy, Dervla, *South from the Limpopo*, John Murray, London, 1997

115 Murphy, Dervla, *The Ukimwi Road: From Kenya to Zimbabwe,* John Murray, London, 1993

115 Murphy, Dervla, 'First, buy your pack animal', *Guardian*, 3 Jan 2009

115 Wroe, Nicholas, 'Free wheeler', *Guardian*, 15 Apr 2006

116 Wheeler, Sara, 'The borrowers', *Telegraph Magazine*, 9 Dec 2006

116 Zakaria, Rafia, *Against White Feminism*, Hamish Hamilton, London 2021

118 Lascelles, Alan, *In Royal Service: The Letters and Journals of Sir Alan Lascelles (1920–1936)*, Hamish Hamilton, London, 1989

119–20 Saro-Wiwa, Noo, *Looking for Transwonderland*, Granta Publications, London, 2012

121 Murphy, Dervla, *Wheels Within Wheels*, John Murray, London, 1979

123 Wheeler, Sara, *Access All Areas: Selected Writings 1990–2010*, Jonathan Cape, London, 2011

128   Burlend, Rebecca, *A True Picture of Emigration*, R. R. Donnelley & Sons Co., Chicago, 1848

128   Kristeva, Julia, 'Motherhood Today'. Accessed: www.kristeva.fr/motherhood.html

128–9 Levy, Deborah, *Things I Don't Want to Know*, Notting Hill Editions, London, 2013

129–30 Painter, Nell Irvin, letter to the *New Yorker*, 16 May 2022

130   Philyaw, Deesha, 'Ain't I a Mommy: Why are so Few Motherhood Memoirs Penned by Women of Color?' 23 Feb 2016. Accessed: www.bitchmedia.org/article/aint-i-a-mommy-0

130   Mellor, Christie, *The Three-Martini Playdate: A Practical Guide to Happy Parenting*, Chronicle, San Francisco, 2004

5 NOT EVEN PAST

131   Whitman, Walt, *Leaves of Grass,* Whitman, 1855

140   Welty, Eudora, *Delta Wedding*, Harcourt, Brace, San Diego, 1946

140   Welty, Eudora, *One Writer's Beginnings,* Harvard University Press, Cambridge (Mass.), 1984

141   Welty, Eudora, *A Curtain of Green*, Doubleday, New York, 1941

141   Porter, Katherine Anne, introduction to Welty, Eudora, *A Curtain of Green*, Doubleday, New York, 1941

141   Roth Pierpont, Claudia, 'A Perfect Lady', the *New Yorker*, 28 Sep 1998

141   Welty, Eudora, 'Why I Live at the P. O.', in *A Curtain of Green*, op. cit.

148   James, Henry, *Essays in London and Elsewhere*, James R. Osgood, London, 1893

149   Shapland, Jenn, *My Autobiography of Carson McCullers*, Tin House Books, Portland, 2020

150   McCullers, Carson, *The Heart is a Lonely Hunter*, Houghton Mifflin & Co., Boston, 1940

155   Whitman, op. cit.

6 ITS OWN MAP

172   Borges, Jorge Luis, trans. James E. Irby, 'The Waiting', in *Labyrinths*, New Directions, New York, 1964

175   Quote attributed to Edward St Aubyn

176   McEwan, Ian, the Q&A, *Guardian*, 18 Aug 2018

179–80 Neruda, Pablo, trans. Hardie St Martin, *Memoirs*, Farrar, Straus & Giroux, New York, 1977

181   Bedford, Sybille, *The Sudden View*, Victor Gollancz, London, 1953

7 THIN PATHS

196–7 Zhang Xinxin and Sang Ye, trans. W. J. F. Jenner & Delia Davin, *Chinese Lives*, Macmillan, London, 1987

197   Raban, Jonathan, *For Love & Money: Writing, Reading, Travelling 1969–87*, HarperCollins, New York, 1989

199–200 Watts, Jonathan, *When a Billion Chinese Jump*, Faber, London, 2011

204   Scott, Danny, 'Matthew McConaughey on his political ambitions and turbulent childhood', *The Times*, 22 Sep 2021

210   Young, Gavin, *Slow Boats to China*, Hutchinson, London, 1981

210 Young, Gavin, *Slow Boats Home*, Hutchinson, London, 1985
211 Raban, Jonathan, *For Love & Money: Writing, Reading, Travelling, 1968–1987*, Collins Harvill, London, 1987
211 White, E. B., 'Across the Street and into the Grill', *New Yorker*, 1950
211 Wheeler, Sara, 'Where have all the female travel writers gone?', *Guardian*, 28 Feb 2017
211 Birkett, Dea, *Serpent in Paradise*, Picador, London, 1997
213 Hawks, Tony, *Round Ireland with a Fridge*, Ebury Publishing, London, 1998
213 Mirrlees, Hope, *Paris: A Poem*, The Hogarth Press, London, 1920
213 Brown, Craig, 'Nothing is real: the slippery art of biography', *The Times Literary Supplement*, 10 Sep 2021
215 Bird, Isabella, *Letters to Henrietta* (ed. Kay Chubbock), John Murray, London, 2002
216–8 Bird, Isabella, *The Yangtze Valley and Beyond*, John Murray, London, 1899
219 Ferguson, Niall, 'Most threatening when weak?: The risks China poses to global security', *The Times Literary Supplement*, 2 Jul 2021
219 Doshi, Rush, *The Long Game: China's Grand Strategy to Displace American Order*, OUP, Oxford 2021

8 DON'T WAKE SOFIA

222 Stuermer, Michael, *Putin and the Rise of Russia*, Weidenfeld & Nicolson, London, 2008
223–4 Chekhov, Anton, 'The Lady with the Little Dog', originally: *Russian Stories*, 1899, 1904
227 Raban, op. cit.
235 Tolstoy, Nikolai, 'Full text of Tolstoy's letters to his wife'. Accessed: https://archive.org/stream/jstor-25108272/25108272_djvu.txt

9 'THE BRONX, NO THONX'

250 Poe, Edgar Allan, letter to Thomas Holley Chivers, 22 Jul 1846, cited https://www.eapoe.org/works/letters/p4607220.htm
251 Amend, Christoph and Diez, Georg, 'I don't know America anymore', translation of an interview with Don DeLillo, by, published in *Die Zeit*, 11 Oct 2007
257 Niarchos, Nicholas, 'La Morada, a Crucible of Resistance', the *New Yorker*, 2 Oct 2017
258 Jonnes, Jill, *South Bronx Rising: The Rise, Fall and Resurrection of an American City*, Fordham University Press, New York, 2002; previous edition: *We're Still Here: The Rise, Fall and Resurrection of the South Bronx*, Little, Brown & Co., New York, 1986
258–9 Doctorow, E. L., *World's Fair*, Random House, New York, 1987
259 Crown, Sarah, 'E. L. Doctorow: "I don't have a style, but the books do",' *Guardian*, 23 Jan 2010
260 Poe, Edgar Allan, 'Annabel Lee', first published in the *New-York Daily Tribune*, 1949
264 *The Times*, 9 Jan 1998
264 Doctorow, E. L., *Billy Bathgate*, Random House, New York, 1989
265 Sotomayor, Sonia, *My Beloved World*, Alfred A. Knopf, New York, 2013
265 Fazlalizadeh, Tatyana, *Stop Telling Women to Smile: Stories of Street*

Harassment and How We're Taking Back Our Power, Seal Press, New
York, 2020

266 Cook, Harry T. and Kaplan, Nathan Julius, *The Borough of the Bronx,
1639-1913: Its Marvelous Development and Historical Surroundings*,
published by the authors, 1913

268 Cooke, Marvel, '"I Was Part of the Bronx Slave Market"', the Daily
Compass, 8 Jan 1950

269 Kristof, Nicholas, 'Life and Death in the "Hot Zone"', *New York Times*,
11 Apr 2020

270 de Freytas-Tamura, Kimiko, 'It's Death Towers', *New York Times*,
26 May 2020

272 le Guin, Ursula, 'A Whitewashed Earthsea: How the Sci Fi Channel
wrecked my books', slate.com, 16 Dec 2004

273 Adichie, op. cit.

10 THE SADDEST PLEASURE

274 Thomsen, Moritz, *The Saddest Pleasure*, The Sumach Press, Toronto, 1990

276 Wordsworth, William, 'Lines Written a Few Miles above Tintern
Abbey', in *Lyrical Ballads with a Few Other Poems* (Wordsworth and
Coleridge), J. & A. Arch, London, 1798

281 Paget, Dan, 'Tanzania: The Authoritarian Landslide', *Journal of
Democracy*, Apr 2021

281–2 'The coronavirus could devastate poor countries', *The Economist*,
26 Mar 2020

284 *The Slave Trade of East Africa*, Church Missionary Society, London, 1888

285 Sulivan, George Lydiard, *Dhow Chasing in Zanzibar Waters and on
the Eastern Coast of Africa*, Samson Low, Marston, Low & Searle,
London, 1873

286 Stanley, Henry cited Jeal, Tim, *Stanley: The Impossible Life of Africa's
Greatest Explorer*, Faber, London, 2007

286 Sulivan, op. cit.

287 *African Tidings*, number 43, May 1893

289 Ruete, Emily, *Memoirs of an Arabian Princess in Zanzibar*, trans. Lionel
Strachey, Doubleday, Page and Co., New York, 1907

293 Dickinson, Emily, 'A Wife – at daybreak I shall be', 1862

294 Globefish Research Programme, Vol 124, Food and Agricultural
Organization of the United Nations, 30 Jan 2019

302 Gurnah, Abdulrazak, *Paradise*, Hamish Hamilton, London, 1994

303 Matthiessen, Peter, *The Tree Where Man was Born*, William Collins,
London, 1972

307 Rachel Carson speaking on the *CBS Reports with Eric Sevareid* episode
'The Silent Spring of Rachel Carson', 3 Apr 1963, documented at www.
rachelcarson.org

307 Lepore, Jill, 'The Lessons of "The Lorax"', the *New Yorker*, 28 Nov 2021

311–2 Thomsen, op. cit.

313 Gellhorn, Martha, *Travels with Myself and Another*, op. cit.

315 bell hooks, 'Moving beyond pain', 9 May 2016. Accessed: www.geledes.
org.br/bell-hooks-offers-complicated-criticism-on-beyonces-lemonade/

315 Savory, op. cit.

316 Boston Women's Health Book Collective, *Our Bodies, Ourselves*, Boston
Women's Health Book Collective, Boston, 1970

316 Greer, Germaine, *The Female Eunuch*, MacGibbon & Kee, London, 1970
316 Als, Hilton, 'Joan Didion, The Art of Nonfiction No. 1', *The Paris Review*, issue 176, Spring 2006

Acknowledgements

My first thanks are to Richard Beswick at Little, Brown. Also to Jon Appleton and his team. These thanks come with sincere gratitude and appreciation. As do my thanks to the following: Mark Collins and Sinan Unel in the Bronx; Julia Baker and Jeff Holbrook in Zanzibar; Lisa Baker, Nell Butler, Phil Kolvin, Paul Mendez and Dr Felix Padel.

Cartoon of Sara Wheeler by Willie Rushton courtesy of *The Oldie*. Image from the *Daily Express* courtesy *Daily Express/* Mirrorpix.

And hey, José – get in touch, man. I've changed my mind. About everything.

Index